The Dividing Wall

including

Beat The Dividing Wall

2 sets of photocopiable worksheets designed to encourage and aid with the learning of division. Extension sheets which can be used with a formal method of short division are also provided.

The sheets are mainly for pupils at upper primary, middle or secondary/high schools.

by

Tony Colledge

ISBN-13: 978-1505378306
ISBN-10: 1505378303

Digital Master: 2.0

For comments, feedback or suggestions you can contact the author via email at

beatthewall@yahoo.com

Other books published by the author include

Beat The Wall

Build The Wall

Pascal's Triangle

Other books to be published in 2015 include

The Fraction Wall

Beat the Deciwall

About the Author

Tony Colledge was educated at Bishopshalt Grammar School in Hillingdon, London. He gained a BSc in Civil Engineering at Swansea University in 1976 and then the following year he trained as a teacher at Gypsy Hill College, Kingston in London, where he received his PGCE.

He spent 37 years in England teaching maths across the whole ability range to pupils aged from 8 to 16. In 1992 he wrote his first book about looking for patterns in Pascal's Triangle.

When he retired in July 2014, he decided to adapt and publish some of his most successful resources in the hope that other teachers, parents and pupils might find them useful.

About the Books

Beat The Wall resources have been published as a set of books each containing photocopiable worksheets based on a particular area of the Numeracy Curriculum. The original worksheets were successfully used to raise pupil achievement, not only in his own classroom but also by other maths teachers working in the same school. They have also been used in lessons observed by Ofsted inspectors and members of the senior management team on many occasions. The lesson gradings in which these resources were used, always resulted in a classification of either good or excellent/outstanding.

Tony Colledge created the first Wall sheets in 1995 whilst teaching several low ability maths classes. He realised, like many other teachers, that the difficulty those pupils had with learning and recalling tables was like a barrier, or *wall*, which hindered their progress in several areas of the curriculum, including division. Of the many resources he wrote at the time, the Wall sheets were the most successful at improving scores. In fact, they were often requested in lessons by the pupils themselves. Weak and able children alike recognised the progress they were making and enjoyed the challenge of trying to beat their previous scores and times.

Since then, he has continued to develop, trial and improve a range of similar sheets and he is convinced that if these resources are used regularly, and as advised, then all pupils will improve their knowledge and understanding of numeracy in the areas covered by the worksheets.

Using the worksheets

The worksheets are **not** designed to be used as an initial teaching resource.

They are more valuable if used as a measure of pupil understanding and application once they have been introduced to the concept of division. Previously, they might have used counters to share equally, looked at or drawn diagrams, been shown one or more written methods and completed a number of consolidation questions.

The Dividing Wall should be used in conjunction with, or after, the Beat The Wall worksheets. In order to do well with division, pupils should have mastered their tables and multiplication facts by knowing both the question and its answer. Those who have achieved this usually score better with the division worksheets and complete them in significantly faster times. Through regular use of the sheets over several months, most pupils can get full marks in a reasonable time. The challenge then becomes to do them faster and faster.

This book contains two sets of structured photocopiable sheets, followed by some additional extension sheets. These can be used over and over again, not just within the same year but also year after year. The repetition enables a teacher to encourage pupils to improve their results, by recognising and praising what they have done well and identifying what still needs to be learnt. This review process allows the setting of simple learning targets either by the teacher or the pupils themselves. By practising their areas of weakness, this should help them to get a higher total or a faster time on the next occasion.

The Dividing Wall sheets test straight forward recall of multiplication facts by giving the answer and one of its factors. This should reinforce the idea that division is simply the inverse of multiplication. The 24 sheets develop by initially using only the facts from the 2 to 10 times tables and then move on by introducing simple mental calculations like $80 \div 4$. After that, questions involving the 11 and 12 times tables are included. The final few sheets are more challenging as they require mental division of numbers up to 1000.

The Beat The Dividing Wall sheets use different missing number variations, initially up to the 10 times table but later sheets include the 11 and 12 times tables as well.

The extension sheets towards the end of this book are designed to help make the transition from mental division to a formal written method for larger numbers. Notes on divisibility rules and division methods follow this section.

The worksheets in my books have been presented in letter size (approx A4) as this makes the numbers clear for pupils to read. However, in order to save photocopying expense it might be possible to reduce each sheet down to A5 and then produce working masters that have 2 sheets to a page so long as your pupils are still able to read the numbers clearly at that size.

I mostly used these resources as a starter as they quickly got the class to settle and focus on practising their numeracy skills. The sheets could also be used as a filler or break between 2 tasks, especially in longer maths lessons. Sometimes finding work for cover lessons can be a problem but once the pupils and staff know the routine associated with these sheets then they can be an appropriate and useful task for part of that lesson. Setting them as a piece of homework is also possible so long as the children can be trusted not to use a calculator to answer all of the questions.

Initially you might need to allow 10 to 20 minutes of a lesson to complete a sheet, reducing this as the children get better. If you choose to time the pupils, how you achieve this is up to you and your classroom management.

When checking answers, several possibilities could be employed. Choosing which one works best depends on how each individual teacher organises their class and classroom resources.

- Copies of the answer sheets could be made available and pupils could self-mark
- Pupils could swap papers and mark their partner's work from the available answer sheets
- Support staff could work with small groups of pupils, monitoring and discussing issues as they arise when going through the answers
- The teacher or support staff could mark all of the sheets and set targets before returning them
- The teacher could read out the answers for a specific worksheet to the whole class at the same time (with or without swapping papers).

An advantage of the last method is that it allows the teacher to interact with the pupils about certain questions as they move through the sheet. Perhaps asking questions like;

- How do we know that 63 divides by 9? (divisibility rule 6 + 3 = 9)
- How can we tell if a large number divides by 5? Or by 3? Or by 6?
- Who can show me a way to do that question?
- Can anyone do it a different way?
- What is the quotient? Or divisor? Or dividend?
- What type of number is 25?
- What is the square root of 49?
- What do we mean by multiple? Or factor? Or factor pair?
- Who can give me an estimate of the answer?
- Who can give me a real-life problem where that question is what we need to work out?
- If I think the answer is 24. Who can tell me why that can't be right?
- Who can give me a 3 digit number that divides exactly by 7?
- 120 divides by 6 so who can tell me another one between 120 and 150?

Once a sheet has been marked, it can be used to identify errors. Staff, or pupils themselves, could circle 3, 4 or perhaps 5 mistakes as targets. Pupils could then take the sheet home and learn the correct answers for next time. This might also be a way to involve parents in the process.

Where a pupil has not yet achieved full marks for a sheet then an individual target of getting *at least* 1 or 2 additional marks next time might be sufficient, especially with those who struggle to make progress. If the target is set too challenging, e.g. more than 5, then the pupil might make some progress but still feel disappointed because they have failed to reach their goal. By setting appropriate achievable targets, it is possible to continually praise pupils for their progress and this encourages them to keep working towards the satisfying achievement of "full marks".

Similar ideas apply to the setting of time targets, if you feel they are appropriate for your class. Completing a sheet 5 or 10 seconds faster is often a reasonable goal. A time target is incredibly useful for those who regularly get full marks. The challenge then becomes to do it as quickly as they can and this engages the most able pupils in the class as they try to get their best *personal* time or battle with others of a similar ability to hold the record for the fastest time.

With practice, most pupils should be able to complete the Wall sheets in around 5 minutes or less but this will depend largely on how accurately and quickly they can recall tables' facts.

What is a good time for the The Dividing Wall sheets? Correctly completing any sheet from 1 to 20 in 2 minutes or less shows a sound recall of their tables and it is even possible for the very best to achieve around 1 minute or under. Sheets 21 to 24 are more challenging and will require more time to complete. A time of less than 3 minutes would be impressive indeed!

A good time to complete Beat The Dividing Wall sheets 1 to 15 would be around 2 minutes each. Sheets 16 to 30 are likely to require slightly longer times but the more able pupils might still finish sometime between 1 and 2 minutes. Anything faster than that is an amazing achievement.

Generally, a good time depends on which sheet is being done, the age and the ability of the pupils doing them. It is up to the professional judgement of the teacher to assess within the context of their class what constitutes a good or an amazing time.

When pupils have done particularly well with their score or time, this could be recognised by giving them a reward of a star, merit, house point or other system that is used within your school.

In conclusion, simply handing out the sheets, giving the class a set time to do the questions and then reading out the answers is not in itself going to make much headway with their division skills. The sheets need to be set regularly, scores and times could be recorded, preferably in the pupils' workbooks or folders so they have easy access to what they did last time and the targets they are trying to achieve next time. The teacher should spend some time after each sheet collecting scores so they can discuss progress made and interact with the pupils about their new targets. This shows that the teacher values the process and it encourages the pupils to do better next time.

I cannot stress highly enough how important the use of targets, praise and encouragement is in helping pupils who struggle to improve their own weak areas in numeracy. The Dividing Wall and Beat The Dividing Wall worksheets, if used regularly and as explained, will help a wide range of children to achieve better scores and better times through greater understanding of the process of division. Using these resources will not only inform teachers about pupil achievement but also they will help struggling pupils to recognise that they have made progress over time and ultimately that should make them feel positive about themselves, about doing maths and about beating that dividing **wall**.

Tony Colledge

Rules for divisibility

When faced with a division question, a very useful mental exercise to do is to check beforehand if the dividend is exactly divisible or if the answer will include a remainder. The extension sheets 3a to 4c provided towards the end of this book can be used to give pupils some practice with this.

There are many different "tricks" for checking whole number divisibility and you may know some other quick methods not covered here but I have included some basic ones to get you started.

Divide by	Rule for divisibility
2	The last digit is even.
3	The sum of the digits is divisible by 3.
4	Half of the number is even. For larger numbers, the last 2 digits are divisible by 4.
5	The last digit is a 5 or 0.
6	The number is even **and** the sum of the digits is divisible by 3.
7	No simple test. (See below).
8	Half of the number is divisible by 4. For larger numbers, the last 3 digits are divisible by 8. (See below).
9	The sum of the digits is divisible by 9.
10	The last digit is a 0.

Divisibility rule for 7

Without a shadow of a doubt, the numeracy properties of 7 are some of the most interesting and unusual. Try dividing any random number by 7, either by a formal method or by using a calculator if more appropriate for the ability of the class, and look closely at the numbers after the decimal point. The decimal digits represent the 'remainder' and fall into 6 repeating patterns of the same 6 digits. When arranged logically by size, here is an example of what happens:

$$8 \div 7 = 1.142857142857142\ldots$$

$$9 \div 7 = 1.285714285714285\ldots$$

$$10 \div 7 = 1.428571428571428\ldots$$

$$11 \div 7 = 1.571428571428571\ldots$$

$$12 \div 7 = 1.714285714285714\ldots$$

$$13 \div 7 = 1.857142857142857\ldots$$

(Note: Try investigating with different divisors. There are other interesting patterns to be found.)

So it is not surprising that a rule for divisibility by 7 is also quite unusual when compared to the other single digit numbers. I am going to give 2 methods for checking divisibility by 7 but be warned that when checking larger numbers the process gets so lengthy that most people agree it is simpler just to go ahead and use a formal short division method to work the answer out.

For these methods we shall first work out how to check if 294 is divisible by 7.

Method 1

- Split the number into 2 parts so the last digit is separate.   29 and 4
- Double the last digit and subtract it from the rest of the number.   29 – 8 = 21
- If this answer is obviously divisible by 7 then the original number 294 is also divisible by 7.
- If it is not clear whether or not the result is divisible by 7 then repeat the process until reaching a definitive result.

Example: Is **3164** divisible by 7?

Method	Working
Split it up	316 and 4
Double the last digit and subtract	316 – 8 = 308
Split it up	30 and 8
Double the last digit and subtract	30 – 16 = 14

14 is a multiple of 7 so 3164 is divisible by 7.

Method 2

This second method uses a different approach to checking for divisibility.

- Multiply the digits by the values shown in the table below.
- Find the sum of these products.
- If the sum is divisible by 7 then so is the original number.

	Thousands			Units			
	H	T	U	H	T	U	
	x 5	x 4	x 6	x 2	x 3	x 1	
Fill in 294 and multiply				2	9	4	
Add the products				4	+ 27	+ 4	= 35

35 is a multiple of 7 so 294 is divisible by 7.

Example: Is **3164** divisible by 7?

	Thousands			Units		
	H	T	U	H	T	U
	x 5	*x 4*	*x 6*	*x 2*	*x 3*	*x 1*
Fill in 3164 and multiply			3	1	6	4
Add the products			18	+ 2	+ 18	+ 4	= 42

42 is a multiple of 7 so 3164 is divisible by 7.

Method 2 (Negative variation)

I think the *"2 3 1"* pattern is really interesting from a mathematical viewpoint. This method is similar to the previous one but requires some multiplying by negative numbers. If the final answer is 0 or a positive / negative multiple of 7 then the number being checked is also a multiple of 7.

Example: Is **3164** divisible by 7?

	Thousands			Units		
	H	T	U	H	T	U
	x -2	*x -3*	*x -1*	*x 2*	*x 3*	*x 1*
Fill in 3164 and multiply			3	1	6	4
Add the products			-3	+ 2	+ 18	+ 4	= 21

21 is a multiple of 7 so 3164 is divisible by 7.

An interesting fact about this method is that the pattern for the 6 columns covering thousands and units repeats itself for even bigger numbers.
To demonstrate this, the example below extends to thousands of millions (*billions*).

Example: Is **17 278 654 498** divisible by 7?

Thousand Millions			Millions			Thousands			Units		
H	T	U	H	T	U	H	T	U	H	T	U
x -2	*x -3*	*x -1*	*x 2*	*x 3*	*x 1*	*x -2*	*x -3*	*x -1*	*x 2*	*x 3*	*x 1*
1	7	2	7	8	6	5	4	4	9	8
- 3	- 7	+ 4	+ 21	+ 8	- 12	- 15	-4	+ 8	+ 27	+ 8	= 35

35 is a multiple of 7 so 17 278 654 498 is divisible by 7.

Divisibility rule for 8

It states in the divisibility table shown previously that if the last 3 digits of a number are divisible by 8 then the whole number is also divisible by 8. I am going to show you a similar method to the ones above for 7 but this time it will check for divisibility by 8. The good news is that we only ever need to check the last 3 digits, not the entire number.

Example: Is **386 136** divisible by 8?

	Units		
	H	**T**	**U**
	x 4	**x 2**	**x 5**
Fill in just 136 and multiply	1	3	6
Add the products	4	+ 6	+ 30	= 40

40 is a multiple of 8 so 386 136 is divisible by 8.

Example: Is **2 593 642** divisible by 8?

	Units		
	H	**T**	**U**
	x 4	**x 2**	**x 5**
Fill in just 642 and multiply	6	4	2
Add the products	24	+ 8	+ 10	= 42

42 is **not** a multiple of 8 so 2 593 642 is **not** exactly divisible by 8.

Division Methods

Of the 4 basic arithmetic processes, division always seems to be the hardest for pupils to grasp and apply. Initially, I think it is vital for them to experience the practical method of physically sharing counters, or whatever else is appropriate, into equal groups. Some of these tasks will divide exactly and some others might have remainders.

At the next stage, pupils can progress to a simple written method which could involve;

- drawing diagrams
- chunking (subtracting known multiples from the total until reaching zero)
- using number lines for smaller numbers
- using empty number lines for larger numbers

However, when they have mastered their tables' facts, I believe the quickest and most efficient method for division of larger numbers is still a formal short division method, sometimes called 'bus stops' or 'bus shelters'.

The method requires using known multiplication facts and calculating remainders. I am sure that most of you will already know how to do short division but for those who are unsure of the method then I will give some examples below.

Example: 7436 ÷ 4

The number after the divide sign (*divisor*) goes outside the 'bus shelter'.
The number being 'shared' (*dividend*) goes inside the 'bus shelter'.

$$4\overline{)7\ 4\ 3\ 6}$$

$7 \div 4 = 1$ remainder 3
Write 1 above the 7 and carry 3 onto the next 4

$$\begin{array}{c}1\\4\overline{)7\ ^34\ 3\ 6}\end{array}$$

$34 \div 4 = 8$ remainder 2
Write 8 above the 34 and carry 2 onto the next 3

$$\begin{array}{c}1\ 8\\4\overline{)7\ ^34\ ^23\ 6}\end{array}$$

$23 \div 4 = 5$ remainder 3
Write 5 above the 23 and carry 3 onto the next 6

$$\begin{array}{c}1\ 8\ 5\\4\overline{)7\ ^34\ ^23\ ^36}\end{array}$$

$36 \div 4 = 9$ with no remainder
Write 9 above the 36. The question is done.

$$\begin{array}{c}1\ 8\ 5\ 9\\4\overline{)7\ ^34\ ^23\ ^36}\end{array}$$

It is good practice to check answers to ensure that no errors have occurred.
Using the principal that division is the inverse of multiplication, work out 4 x 1859 to see if the product is 7436.

Example: 16254 ÷ 6

The number after the divide sign goes outside the 'bus shelter'.
The number being 'shared' goes inside the 'bus shelter'.

$$6\,\overline{)1\ 6\ 2\ 5\ 4}$$

1 ÷ 6 = 0 remainder 1
Do not write 0 above the 1 but carry 1 onto the next 6

$$6\,\overline{)1\ {}^16\ 2\ 5\ 4}$$

16 ÷ 6 = 2 remainder 4
Write 2 above the 16 and carry 4 onto the next 2

$$\overset{2}{6\,\overline{)1\ {}^16\ {}^42\ 5\ 4}}$$

42 ÷ 6 = 7 remainder 0
Write 7 above the 42 and there is no remainder to carry.

$$\overset{2\ 7}{6\,\overline{)1\ {}^16\ {}^42\ 5\ 4}}$$

5 ÷ 6 = 0 remainder 5
Write 0 above the 5 and carry 5 onto the next 4.

$$\overset{2\ 7\ 0}{6\,\overline{)1\ {}^16\ {}^42\ 5\ {}^54}}$$

54 ÷ 6 = 9 with no remainder.
Write 9 above the 54. The question is done.

$$\overset{2\ 7\ 0\ 9}{6\,\overline{)1\ {}^16\ {}^42\ 5\ {}^54}}$$

Once again work out 6 x 2709 to check if the product is 16254.

Initially, when teaching this method, I would go through a number of worked examples on the board with the whole class, gradually getting them to do more and more of the parts themselves until many of them were confident enough and raring to have a go at some on their own. Extension worksheets 3a, 3b and 3c cover dividends in the hundreds and worksheets 4a, 4b and 4c extend into the thousands. Whilst the more confident pupils were answering those questions, I had the opportunity to work with those who needed additional support with grasping the process.

Referring to the example above, the main errors that often arose in the first few lessons of teaching short division were;

- Carrying 6 or 5 instead of 1 (stage 1).

- Basic errors using multiplication facts or mental subtraction (all stages).
- Transposing the numbers 2 rem 4 and putting 4 rem 2 instead (stage 2).
- Forgetting to put a zero in the middle of the answer when needed (stage 4).
- Swapping the order by seeing 5 ÷ 6 as 6 ÷ 5 and getting 1 remainder 1 (stage 4).

Being aware of the most common faults can be very useful when doing the worked examples as it allows the teacher to identify and prevent possible misunderstandings right from the start.

This formal short division method might not suit all pupils but those who do understand it will be able to do division with very long numbers quite easily. Extension worksheets 5a and 5b are provided to challenge those pupils in particular. Frequently, the children who had completed the extension sheets kept asking for additional longer and longer questions. There is no doubting the fact that they were gaining a huge sense of satisfaction from finally being successful at division.

In conclusion, it would be appropriate to mention something about long division. To be honest, I never bothered teaching a long division method, unless exceptional circumstances required it. Although some of you might disagree with me, I found that long division was an over-complicated process which often resulted in confusion not clarity. However, where some pupils found it difficult to hold more than one piece of information in their head at a time then writing down the working out as they went along was a good strategy that often helped them to achieve a result. In this situation, using any suitable 'customised' method that worked for them was the over-riding key issue.

I think that cultivating success and confidence with numeracy should be the constant aim of all maths teachers. Different methods can be tried but those that cause confusion, and a sense of failure, should never be rigidly imposed upon children. Instead, I believe it is essential that teachers have a variety of methods 'up their sleeves' so all of their pupils can choose and use a suitable method they both understand and are ready for. In this way we can help children to overcome their doubts and fears and become as proficient at division as they are in other areas of the curriculum.

The Dividing Wall

Sheets 1 to 10 test recall of multiplication facts when dividing by 2 to 10

Sheets 11 to 15 include some additional questions like 80 ÷ 4

Sheets 16 to 20 test recall of multiplication facts when dividing by 2 to 12

Sheets 21 to 24 test mental division of numbers up to 1000 by 2 to 12

Answers are provided at the end of this section.

An example of a pupil record sheet is provided with the answers.

NAME: _____   DATE: _____

THE DIVIDING WALL

KEY FOCUS: *Dividing by 2 to 10*

How many can you get right ?

How fast can you do it ?

What will you try to get next time ?

YOU CAN DO IT !

SCORE	37
mins	**secs**
TARGETS	37
mins	**secs**

1

$4 \div 2 =$

$6 \div 3 =$ $8 \div 2 =$

$9 \div 3 =$ $10 \div 5 =$ $12 \div 4 =$

$14 \div 7 =$ $15 \div 3 =$ $16 \div 4 =$ $18 \div 6 =$

$20 \div 5 =$ $21 \div 3 =$ $24 \div 4 =$ $25 \div 5 =$ $27 \div 9 =$

$28 \div 7 =$ $30 \div 5 =$ $32 \div 4 =$ $35 \div 7 =$

$36 \div 6 =$ $40 \div 5 =$ $42 \div 6 =$ $45 \div 9 =$ $48 \div 6 =$

$49 \div 7 =$ $50 \div 10 =$ $54 \div 6 =$ $56 \div 7 =$

$60 \div 10 =$ $63 \div 9 =$ $64 \div 8 =$ $70 \div 7 =$ $72 \div 8 =$

$80 \div 8 =$ $81 \div 9 =$ $90 \div 10 =$ $100 \div 10 =$

NAME: _____   DATE: _____

THE DIVIDING WALL

KEY FOCUS: *Dividing by 2 to 10*

How many can you get right ?

How fast can you do it ?

What will you try to get next time ?

YOU CAN DO IT !

SCORE	/ 37
mins	secs

TARGETS	/ 37
mins	secs

$100 \div 10 =$

$81 \div 9 =$ $90 \div 9 =$

$70 \div 7 =$ $72 \div 9 =$ $80 \div 10 =$

$56 \div 8 =$ $60 \div 10 =$ $63 \div 7 =$ $64 \div 8 =$

$45 \div 9 =$ $48 \div 8 =$ $49 \div 7 =$ $50 \div 10 =$ $54 \div 9 =$

$35 \div 5 =$ $36 \div 4 =$ $40 \div 8 =$ $42 \div 7 =$

$25 \div 5 =$ $27 \div 3 =$ $28 \div 7 =$ $30 \div 5 =$ $32 \div 8 =$

$18 \div 6 =$ $20 \div 4 =$ $21 \div 3 =$ $24 \div 3 =$

$10 \div 2 =$ $12 \div 3 =$ $14 \div 7 =$ $15 \div 5 =$ $16 \div 4 =$

$4 \div 2 =$ $6 \div 2 =$ $8 \div 4 =$ $9 \div 3 =$

THE DIVIDING WALL

NAME: _____

DATE: _____

3

KEY FOCUS: *Dividing by 2 to 10*

How many can you get right ?

How fast can you do it ?

What will you try to get next time ?

YOU CAN DO IT !

SCORE	/ 37
mins	secs
TARGETS	/ 37
mins	secs

$21 \div 3 =$

$36 \div 6 =$ $15 \div 3 =$

$14 \div 2 =$ $35 \div 7 =$ $54 \div 6 =$

$32 \div 4 =$ $72 \div 9 =$ $6 \div 2 =$ $40 \div 5 =$

$100 \div 10 =$ $27 \div 9 =$ $18 \div 2 =$ $63 \div 9 =$ $60 \div 6 =$

$56 \div 7 =$ $20 \div 5 =$ $42 \div 6 =$ $10 \div 5 =$

$81 \div 9 =$ $9 \div 3 =$ $50 \div 10 =$ $64 \div 8 =$ $24 \div 3 =$

$30 \div 5 =$ $49 \div 7 =$ $4 \div 2 =$ $80 \div 8 =$

$28 \div 7 =$ $45 \div 9 =$ $16 \div 2 =$ $90 \div 10 =$ $12 \div 6 =$

$8 \div 2 =$ $70 \div 7 =$ $25 \div 5 =$ $48 \div 8 =$

NAME: _____

DATE: _____

THE DIVIDING WALL

4

KEY FOCUS: *Dividing by 2 to 10*

How many can you get right ?

How fast can you do it ?

What will you try to get next time ?

YOU CAN DO IT !

SCORE	37
mins	secs
TARGETS	37
mins	secs

$42 \div 7 =$

$54 \div 6 =$ $16 \div 4 =$

$16 \div 8 =$ $49 \div 7 =$ $70 \div 10 =$

$36 \div 6 =$ $40 \div 4 =$ $63 \div 7 =$ $15 \div 5 =$

$45 \div 5 =$ $14 \div 7 =$ $90 \div 9 =$ $12 \div 4 =$ $56 \div 8 =$

$30 \div 6 =$ $28 \div 4 =$ $72 \div 8 =$ $9 \div 3 =$

$25 \div 5 =$ $12 \div 6 =$ $36 \div 9 =$ $40 \div 8 =$ $100 \div 10 =$

$32 \div 8 =$ $18 \div 6 =$ $50 \div 5 =$ $24 \div 4 =$

$80 \div 10 =$ $48 \div 6 =$ $35 \div 7 =$ $81 \div 9 =$ $20 \div 4 =$

$18 \div 2 =$ $24 \div 8 =$ $60 \div 6 =$ $27 \div 3 =$

NAME: _____

DATE: _____

THE DIVIDING WALL

KEY FOCUS: *Dividing by 2 to 10*

How many can you get right ?

How fast can you do it ?

What will you try to get next time ?

YOU CAN DO IT !

SCORE	/37
mins	secs
TARGETS	/37
mins	secs

$56 \div 7 =$

$36 \div 6 =$ $63 \div 9 =$

$40 \div 5 =$ $60 \div 10 =$ $64 \div 8 =$

$24 \div 3 =$ $54 \div 9 =$ $25 \div 5 =$ $80 \div 8 =$

$18 \div 9 =$ $12 \div 3 =$ $72 \div 9 =$ $42 \div 6 =$ $18 \div 3 =$

$32 \div 4 =$ $40 \div 10 =$ $81 \div 9 =$ $28 \div 7 =$

$70 \div 7 =$ $16 \div 4 =$ $9 \div 3 =$ $45 \div 9 =$ $50 \div 10 =$

$24 \div 6 =$ $27 \div 9 =$ $35 \div 5 =$ $12 \div 6 =$

$48 \div 8 =$ $15 \div 3 =$ $36 \div 4 =$ $30 \div 5 =$ $100 \div 10 =$

$20 \div 5 =$ $49 \div 7 =$ $90 \div 10 =$ $16 \div 2 =$

THE DIVIDING WALL

6

NAME: _____

DATE: _____

KEY FOCUS: *Dividing by 2 to 10*

How many can you get right ?

How fast can you do it ?

What will you try to get next time ?

YOU CAN DO IT !

SCORE	/42
mins	secs
TARGETS	/42
mins	secs

$24 \div 4 =$
$24 \div 3 =$

$42 \div 7 =$ $10 \div 5 =$

$8 \div 4 =$ $100 \div 10 =$ $15 \div 3 =$

$64 \div 8 =$ $16 \div 4 =$ $36 \div 6 =$ $21 \div 3 =$
$16 \div 2 =$ $36 \div 4 =$

$45 \div 9 =$ $80 \div 8 =$ $27 \div 9 =$ $72 \div 8 =$ $14 \div 7 =$

$6 \div 3 =$ $54 \div 6 =$ $90 \div 10 =$ $35 \div 7 =$

$32 \div 4 =$ $12 \div 4 =$ $81 \div 9 =$ $40 \div 5 =$ $30 \div 5 =$
$12 \div 6 =$ $40 \div 4 =$

$60 \div 6 =$ $49 \div 7 =$ $9 \div 3 =$ $56 \div 8 =$

$20 \div 5 =$ $48 \div 6 =$ $50 \div 10 =$ $28 \div 7 =$ $25 \div 5 =$

$18 \div 2 =$ $70 \div 7 =$ $18 \div 6 =$ $4 \div 2 =$

NAME: _____   DATE: _____

THE DIVIDING WALL

7

KEY FOCUS: *Dividing by 2 to 10*

How many can you get right ?

How fast can you do it ?

What will you try to get next time ?

YOU CAN DO IT !

SCORE	/42
mins	secs
TARGETS	/42
mins	secs

$48 \div 8 =$

$25 \div 5 =$ $81 \div 9 =$

$100 \div 10 =$ $18 \div 9 =$ $56 \div 8 =$

$49 \div 7 =$ $30 \div 5 =$ $63 \div 9 =$ $54 \div 6 =$

$14 \div 7 =$ $36 \div 6 =$ $36 \div 9 =$ $8 \div 4 =$ $12 \div 3 =$ $12 \div 2 =$ $70 \div 7 =$

$42 \div 6 =$ $50 \div 10 =$ $45 \div 9 =$ $21 \div 3 =$

$18 \div 3 =$ $72 \div 8 =$ $40 \div 8 =$ $40 \div 4 =$ $35 \div 7 =$ $80 \div 8 =$

$90 \div 10 =$ $32 \div 4 =$ $15 \div 5 =$ $28 \div 4 =$

$6 \div 3 =$ $24 \div 6 =$ $24 \div 8 =$ $10 \div 5 =$ $16 \div 4 =$ $16 \div 8 =$ $9 \div 3 =$

$20 \div 4 =$ $64 \div 8 =$ $27 \div 9 =$ $60 \div 6 =$

NAME: _____   DATE: _____

THE DIVIDING WALL

8

KEY FOCUS: *Dividing by 2 to 10*

How many can you get right ?

How fast can you do it ?

What will you try to get next time ?

YOU CAN DO IT !

SCORE	/45
mins	secs

TARGETS	/45
mins	secs

$81 \div 9 =$

$25 \div 5 =$ $63 \div 7 =$

$72 \div 9 =$ $50 \div 5 =$ $32 \div 8 =$

$4 \div 2 =$ $20 \div 2 =$ / $20 \div 5 =$ $36 \div 4 =$ / $36 \div 6 =$ $90 \div 10 =$

$6 \div 2 =$ $56 \div 7 =$ $10 \div 2 =$ $48 \div 6 =$ $60 \div 10 =$

$28 \div 7 =$ $12 \div 4 =$ / $12 \div 6 =$ $30 \div 3 =$ / $30 \div 5 =$ $42 \div 7 =$

$35 \div 5 =$ $27 \div 3 =$ $100 \div 10 =$ $9 \div 3 =$ $54 \div 9 =$

$14 \div 2 =$ $40 \div 5 =$ / $40 \div 4 =$ $16 \div 2 =$ / $16 \div 4 =$ $15 \div 3 =$

$49 \div 7 =$ $8 \div 2 =$ $64 \div 8 =$ $70 \div 10 =$ $45 \div 5 =$

$80 \div 8 =$ $18 \div 3 =$ / $18 \div 2 =$ $24 \div 4 =$ / $24 \div 3 =$ $21 \div 7 =$

NAME: _____   **DATE:** _____

THE DIVIDING WALL

KEY FOCUS: *Dividing by 2 to 10*

How many can you get right ?

How fast can you do it ?

What will you try to get next time ?

YOU CAN DO IT !

SCORE	/ 45
mins	secs
TARGETS	/ 45
mins	secs

$32 \div 4 =$

$64 \div 8 =$ $49 \div 7 =$

$70 \div 7 =$ $45 \div 9 =$ $9 \div 3 =$

$21 \div 3 =$ $54 \div 6 =$ $14 \div 7 =$ $72 \div 8 =$

$80 \div 10 =$ $28 \div 4 =$ $6 \div 3 =$ $42 \div 6 =$ $81 \div 9 =$

$30 \div 3 =$ $16 \div 4 =$ $36 \div 6 =$ $20 \div 10 =$
$30 \div 5 =$ $16 \div 8 =$ $36 \div 9 =$ $20 \div 4 =$

$50 \div 10 =$ $27 \div 9 =$ $48 \div 8 =$ $4 \div 2 =$ $60 \div 6 =$

$24 \div 6 =$ $40 \div 8 =$ $12 \div 2 =$ $18 \div 6 =$
$24 \div 8 =$ $40 \div 10 =$ $12 \div 3 =$ $18 \div 9 =$

$90 \div 9 =$ $10 \div 5 =$ $56 \div 8 =$ $15 \div 5 =$ $100 \div 10 =$

$25 \div 5 =$ $63 \div 9 =$ $8 \div 4 =$ $35 \div 7 =$

THE DIVIDING WALL

10

KEY FOCUS: *Dividing by 2 to 10*

How many can you get right ?

How fast can you do it ?

What will you try to get next time ?

YOU CAN DO IT !

SCORE	/45
mins	secs
TARGETS	/45
mins	secs

$63 \div 7 =$

$56 \div 7 =$ $81 \div 9 =$

$10 \div 2 =$ $64 \div 8 =$ $35 \div 5 =$

$12 \div 6 =$
$12 \div 4 =$ $28 \div 7 =$ $25 \div 5 =$ $30 \div 10 =$
$30 \div 6 =$

$6 \div 2 =$ $90 \div 10 =$ $49 \div 7 =$ $72 \div 9 =$ $50 \div 5 =$

$20 \div 2 =$
$20 \div 5 =$ $48 \div 6 =$ $14 \div 2 =$ $18 \div 3 =$
$18 \div 2 =$

$21 \div 7 =$ $42 \div 7 =$ $100 \div 10 =$ $8 \div 2 =$ $32 \div 8 =$

$36 \div 4 =$
$36 \div 6 =$ $9 \div 3 =$ $70 \div 10 =$ $24 \div 3 =$
$24 \div 4 =$

$4 \div 2 =$ $54 \div 9 =$ $15 \div 3 =$ $80 \div 8 =$ $27 \div 3 =$

$16 \div 2 =$
$16 \div 4 =$ $60 \div 10 =$ $45 \div 5 =$ $40 \div 8 =$
$40 \div 10 =$

NAME: _____   DATE: _____

THE DIVIDING WALL

11

KEY FOCUS: *Dividing by 2 to 10*

How many can you get right ?

How fast can you do it ?

What will you try to get next time ?

YOU CAN DO IT !

SCORE	/50
mins	secs
TARGETS	/50
mins	secs

$54 \div 6 =$

$21 \div 3 =$ $45 \div 9 =$

$42 \div 6 =$ $32 \div 4 =$ / $32 \div 2 =$ $25 \div 5 =$

$35 \div 7 =$ $40 \div 5 =$ / $40 \div 4 =$ $18 \div 2 =$ / $18 \div 3 =$ $72 \div 8 =$

$90 \div 9 =$ $16 \div 4 =$ / $16 \div 2 =$ $4 \div 2 =$ $48 \div 6 =$ / $48 \div 2 =$ $15 \div 3 =$

$30 \div 5 =$ / $30 \div 3 =$ $27 \div 9 =$ $64 \div 8 =$ $20 \div 2 =$ / $20 \div 5 =$

$24 \div 4 =$ / $24 \div 3 =$ $81 \div 9 =$ $49 \div 7 =$ $6 \div 3 =$ $36 \div 6 =$ / $36 \div 4 =$

$70 \div 7 =$ / $70 \div 2 =$ $9 \div 3 =$ $63 \div 9 =$ $80 \div 10 =$ / $80 \div 4 =$

$56 \div 7 =$ $12 \div 4 =$ / $12 \div 6 =$ $8 \div 2 =$ $60 \div 10 =$ / $60 \div 3 =$ $100 \div 10 =$

$14 \div 2 =$ $50 \div 10 =$ $28 \div 7 =$ $10 \div 5 =$

NAME: _____   DATE: _____

THE DIVIDING WALL

12

KEY FOCUS: *Dividing by 2 to 10*

How many can you get right ?

How fast can you do it ?

What will you try to get next time ?

YOU CAN DO IT !

SCORE	/50
mins	secs
TARGETS	/50
mins	secs

$36 \div 9 =$
$36 \div 6 =$

$70 \div 10 =$
$70 \div 2 =$

$12 \div 3 =$
$12 \div 2 =$

$16 \div 8 =$
$16 \div 4 =$

$8 \div 2 =$

$30 \div 10 =$
$30 \div 6 =$

$40 \div 10 =$
$40 \div 8 =$

$64 \div 8 =$

$42 \div 7 =$

$18 \div 9 =$
$18 \div 6 =$

$32 \div 8 =$
$32 \div 2 =$

$50 \div 5 =$

$6 \div 2 =$

$56 \div 8 =$

$80 \div 8 =$
$80 \div 4 =$

$20 \div 10 =$
$20 \div 4 =$

$27 \div 3 =$

$90 \div 10 =$

$24 \div 6 =$
$24 \div 8 =$

$49 \div 7 =$

$48 \div 8 =$
$48 \div 2 =$

$9 \div 3 =$

$60 \div 10 =$
$60 \div 3 =$

$35 \div 5 =$

$10 \div 5 =$

$72 \div 9 =$

$63 \div 7 =$

$14 \div 2 =$

$25 \div 5 =$

$54 \div 9 =$

$4 \div 2 =$

$100 \div 10 =$

$21 \div 7 =$

$45 \div 9 =$

$15 \div 5 =$

$28 \div 4 =$

$81 \div 9 =$

NAME: _____   DATE: _____

THE DIVIDING WALL

13

KEY FOCUS: *Dividing by 2 to 10*

How many can you get right ?

How fast can you do it ?

What will you try to get next time ?

YOU CAN DO IT !

SCORE	/50
mins	secs
TARGETS	/50
mins	secs

$18 \div 2 =$
$18 \div 3 =$

$50 \div 10 =$
$50 \div 2 =$

$36 \div 6 =$
$36 \div 4 =$

$32 \div 4 =$
$32 \div 2 =$

$72 \div 8 =$

$30 \div 5 =$
$30 \div 3 =$

$90 \div 9 =$
$90 \div 3 =$

$10 \div 5 =$

$56 \div 7 =$

$12 \div 4 =$
$12 \div 6 =$

$64 \div 8 =$

$27 \div 9 =$

$100 \div 10 =$

$9 \div 3 =$

$49 \div 7 =$

$16 \div 2 =$
$16 \div 4 =$

$14 \div 7 =$

$54 \div 6 =$

$20 \div 2 =$
$20 \div 5 =$

$63 \div 9 =$

$28 \div 7 =$

$60 \div 10 =$

$25 \div 5 =$

$6 \div 3 =$

$40 \div 4 =$
$40 \div 5 =$

$4 \div 2 =$

$45 \div 9 =$

$24 \div 3 =$
$24 \div 4 =$

$21 \div 3 =$

$70 \div 7 =$

$8 \div 2 =$

$81 \div 9 =$

$35 \div 7 =$

$48 \div 6 =$
$48 \div 2 =$

$15 \div 3 =$

$42 \div 6 =$

$80 \div 10 =$
$80 \div 4 =$

THE DIVIDING WALL

14

KEY FOCUS: *Dividing by 2 to 10*

How many can you get right ?

How fast can you do it ?

What will you try to get next time ?

YOU CAN DO IT !

SCORE	/50
mins	secs

TARGETS	/50
mins	secs

$54 \div 9 =$

$32 \div 8 =$ $21 \div 7 =$

$10 \div 2 =$ $45 \div 5 =$ $72 \div 9 =$

$36 \div 9 =$
$36 \div 6 =$ $64 \div 8 =$ $15 \div 5 =$ $28 \div 7 =$
$28 \div 2 =$

$27 \div 3 =$ $30 \div 6 =$
$30 \div 2 =$ $81 \div 9 =$ $12 \div 3 =$
$12 \div 2 =$ $25 \div 5 =$

$56 \div 8 =$ $16 \div 4 =$
$16 \div 8 =$ $40 \div 10 =$
$40 \div 8 =$ $6 \div 2 =$

$50 \div 10 =$ $4 \div 2 =$ $48 \div 6 =$
$48 \div 4 =$ $49 \div 7 =$ $90 \div 9 =$

$35 \div 5 =$ $18 \div 9 =$
$18 \div 6 =$ $100 \div 4 =$
$100 \div 5 =$ $14 \div 2 =$

$8 \div 4 =$ $60 \div 10 =$
$60 \div 3 =$ $42 \div 7 =$ $24 \div 6 =$
$24 \div 8 =$ $80 \div 8 =$

$20 \div 10 =$
$20 \div 4 =$ $63 \div 9 =$ $9 \div 3 =$ $70 \div 7 =$
$70 \div 2 =$

NAME: _____   DATE: _____

THE DIVIDING WALL

15

KEY FOCUS: *Dividing by 2 to 10*

How many can you get right ?

How fast can you do it ?

What will you try to get next time ?

YOU CAN DO IT !

SCORE	/50
mins	secs

TARGETS	/50
mins	secs

$40 \div 5 =$
$40 \div 4 =$

$27 \div 9 =$ $56 \div 7 =$

$72 \div 8 =$ $18 \div 3 =$
$18 \div 2 =$ $42 \div 6 =$

$12 \div 6 =$ $63 \div 7 =$ $32 \div 4 =$ $50 \div 10 =$
$12 \div 4 =$ $50 \div 2 =$

$10 \div 5 =$ $49 \div 7 =$ $90 \div 9 =$ $64 \div 8 =$ $9 \div 3 =$
                                    $90 \div 2 =$

$36 \div 6 =$ $6 \div 3 =$ $81 \div 9 =$ $20 \div 2 =$
$36 \div 9 =$ $20 \div 5 =$

$15 \div 3 =$ $70 \div 7 =$ $24 \div 3 =$ $4 \div 2 =$ $25 \div 5 =$
                                    $24 \div 4 =$

$30 \div 5 =$ $45 \div 9 =$ $14 \div 7 =$ $48 \div 6 =$
$30 \div 3 =$ $48 \div 4 =$

$100 \div 4 =$ $8 \div 2 =$ $60 \div 10 =$ $35 \div 7 =$ $21 \div 3 =$
                                    $60 \div 4 =$

$16 \div 2 =$ $28 \div 7 =$ $54 \div 6 =$ $80 \div 10 =$
$16 \div 4 =$ $80 \div 4 =$

NAME: _____   DATE: _____

THE DIVIDING WALL

KEY FOCUS: *Dividing by 2 to 12*

How many can you get right ?

How fast can you do it ?

What will you try to get next time ?

YOU CAN DO IT !

16

SCORE	/37
mins	secs
TARGETS	/37
mins	secs

$54 \div 6 =$

$72 \div 9 =$ $36 \div 3 =$

$121 \div 11 =$ $24 \div 8 =$ $144 \div 12 =$

$120 \div 12 =$ $49 \div 7 =$ $55 \div 11 =$ $36 \div 4 =$

$27 \div 9 =$ $48 \div 4 =$ $21 \div 7 =$ $64 \div 8 =$ $100 \div 5 =$

$28 \div 4 =$ $132 \div 12 =$ $60 \div 3 =$ $45 \div 9 =$

$42 \div 7 =$ $24 \div 2 =$ $63 \div 7 =$ $99 \div 9 =$ $36 \div 6 =$

$56 \div 8 =$ $30 \div 5 =$ $44 \div 4 =$ $108 \div 12 =$

$48 \div 6 =$ $25 \div 5 =$ $110 \div 11 =$ $18 \div 3 =$ $32 \div 8 =$

$40 \div 8 =$ $81 \div 9 =$ $24 \div 6 =$ $60 \div 12 =$

NAME: _____   DATE: _____

THE DIVIDING WALL

17

KEY FOCUS: *Dividing by 2 to 12*

How many can you get right ?

How fast can you do it ?

What will you try to get next time ?

YOU CAN DO IT !

SCORE	/37
mins	secs
TARGETS	/37
mins	secs

$108 \div 9 =$

$33 \div 3 =$ $56 \div 7 =$

$81 \div 9 =$ $48 \div 8 =$ $24 \div 12 =$

$32 \div 4 =$ $88 \div 8 =$ $63 \div 9 =$ $30 \div 6 =$

$49 \div 7 =$ $60 \div 5 =$ $21 \div 3 =$ $132 \div 11 =$ $24 \div 4 =$

$36 \div 9 =$ $121 \div 11 =$ $45 \div 5 =$ $80 \div 4 =$

$24 \div 3 =$ $54 \div 6 =$ $60 \div 2 =$ $72 \div 8 =$ $36 \div 12 =$

$33 \div 11 =$ $144 \div 12 =$ $27 \div 3 =$ $40 \div 5 =$

$42 \div 6 =$ $35 \div 7 =$ $64 \div 8 =$ $25 \div 5 =$ $28 \div 7 =$

$36 \div 6 =$ $48 \div 12 =$ $110 \div 10 =$ $100 \div 4 =$

THE DIVIDING WALL

18

KEY FOCUS: *Dividing by 2 to 12*

How many can you get right ?

How fast can you do it ?

What will you try to get next time ?

YOU CAN DO IT !

SCORE	/37
mins	secs

TARGETS	/37
mins	secs

$63 \div 7 =$

$144 \div 12 =$ $40 \div 8 =$

$60 \div 12 =$ $24 \div 8 =$ $99 \div 11 =$

$36 \div 3 =$ $35 \div 5 =$ $28 \div 4 =$ $64 \div 8 =$

$56 \div 8 =$ $21 \div 7 =$ $81 \div 9 =$ $42 \div 7 =$ $132 \div 12 =$

$32 \div 8 =$ $110 \div 11 =$ $18 \div 3 =$ $27 \div 9 =$

$48 \div 4 =$ $25 \div 5 =$ $108 \div 12 =$ $36 \div 4 =$ $60 \div 4 =$

$45 \div 9 =$ $80 \div 4 =$ $24 \div 2 =$ $121 \div 11 =$

$77 \div 7 =$ $36 \div 6 =$ $54 \div 9 =$ $30 \div 5 =$ $48 \div 6 =$

$24 \div 6 =$ $120 \div 12 =$ $72 \div 9 =$ $49 \div 7 =$

THE DIVIDING WALL

19

KEY FOCUS: *Dividing by 2 to 12*

How many can you get right ?

How fast can you do it ?

What will you try to get next time ?

YOU CAN DO IT !

SCORE	/37
mins	secs
TARGETS	/37
mins	secs

$64 \div 8 =$

$56 \div 7 =$ $36 \div 9 =$

$100 \div 5 =$ $49 \div 7 =$ $90 \div 3 =$

$42 \div 6 =$ $60 \div 5 =$ $24 \div 4 =$ $72 \div 8 =$

$27 \div 3 =$ $144 \div 12 =$ $33 \div 3 =$ $63 \div 9 =$ $32 \div 4 =$

$108 \div 9 =$ $24 \div 12 =$ $121 \div 11 =$ $70 \div 2 =$

$54 \div 6 =$ $36 \div 12 =$ $81 \div 9 =$ $16 \div 4 =$ $40 \div 5 =$

$25 \div 5 =$ $48 \div 8 =$ $21 \div 3 =$ $110 \div 10 =$

$28 \div 7 =$ $132 \div 11 =$ $18 \div 6 =$ $45 \div 5 =$ $30 \div 6 =$

$36 \div 6 =$ $24 \div 3 =$ $48 \div 12 =$ $35 \div 7 =$

NAME: _____   DATE: _____

THE DIVIDING WALL

KEY FOCUS: *Dividing by 2 to 12*

How many can you get right ?

How fast can you do it ?

What will you try to get next time ?

YOU CAN DO IT !

20

SCORE	/37
mins	secs
TARGETS	/37
mins	secs

$60 \div 12 =$

$48 \div 4 =$ $54 \div 9 =$

$63 \div 7 =$ $24 \div 2 =$ $36 \div 6 =$

$88 \div 11 =$ $56 \div 8 =$ $132 \div 12 =$ $25 \div 5 =$

$32 \div 8 =$ $108 \div 12 =$ $36 \div 3 =$ $48 \div 6 =$ $55 \div 5 =$

$100 \div 4 =$ $121 \div 11 =$ $24 \div 6 =$ $42 \div 7 =$

$144 \div 12 =$ $27 \div 9 =$ $36 \div 4 =$ $64 \div 8 =$ $18 \div 3 =$

$45 \div 9 =$ $60 \div 3 =$ $81 \div 9 =$ $21 \div 7 =$

$120 \div 4 =$ $72 \div 9 =$ $24 \div 8 =$ $110 \div 11 =$ $35 \div 5 =$

$40 \div 8 =$ $28 \div 4 =$ $49 \div 7 =$ $30 \div 5 =$

THE DIVIDING WALL

NAME: _____   DATE: _____

21

KEY FOCUS: *Mental division of numbers up to 1000 by 2 to 12*

How many can you get right ?

How fast can you do it ?

What will you try to get next time ?

YOU CAN DO IT !

SCORE	/37
mins	secs
TARGETS	/37
mins	secs

$420 \div 6 =$

$240 \div 12 =$ $280 \div 7 =$

$540 \div 9 =$ $160 \div 8 =$ $360 \div 6 =$

$220 \div 11 =$ $490 \div 7 =$ $640 \div 8 =$ $250 \div 5 =$

$480 \div 6 =$ $450 \div 9 =$ $240 \div 3 =$ $180 \div 2 =$ $350 \div 7 =$

$160 \div 8 =$ $630 \div 7 =$ $360 \div 9 =$ $720 \div 8 =$

$240 \div 6 =$ $350 \div 5 =$ $300 \div 6 =$ $320 \div 4 =$ $120 \div 3 =$

$360 \div 12 =$ $480 \div 8 =$ $560 \div 7 =$ $400 \div 8 =$

$150 \div 3 =$ $240 \div 8 =$ $160 \div 4 =$ $630 \div 9 =$ $540 \div 6 =$

$180 \div 3 =$ $270 \div 9 =$ $360 \div 4 =$ $90 \div 3 =$

THE DIVIDING WALL

KEY FOCUS: *Mental division of numbers up to 1000 by 2 to 12*

NAME: _____   DATE: _____

22

How many can you get right ?

How fast can you do it ?

What will you try to get next time ?

YOU CAN DO IT !

SCORE	/37
mins	secs
TARGETS	/37
mins	secs

$120 \div 5 =$

$168 \div 8 =$ $132 \div 11 =$

$108 \div 4 =$ $720 \div 9 =$ $150 \div 3 =$

$217 \div 7 =$ $110 \div 2 =$ $170 \div 5 =$ $426 \div 6 =$

$770 \div 11 =$ $240 \div 8 =$ $190 \div 2 =$ $180 \div 5 =$ $249 \div 3 =$

$90 \div 6 =$ $130 \div 2 =$ $819 \div 9 =$ $350 \div 7 =$

$180 \div 4 =$ $120 \div 6 =$ $186 \div 3 =$ $96 \div 12 =$ $116 \div 2 =$

$240 \div 8 =$ $98 \div 2 =$ $360 \div 12 =$ $450 \div 9 =$

$630 \div 7 =$ $140 \div 7 =$ $540 \div 9 =$ $120 \div 4 =$ $330 \div 11 =$

$105 \div 5 =$ $150 \div 2 =$ $216 \div 3 =$ $189 \div 9 =$

NAME: _____   DATE: _____

THE DIVIDING WALL

KEY FOCUS: *Mental division of numbers up to 1000 by 2 to 12*

How many can you get right ?

How fast can you do it ?

What will you try to get next time ?

YOU CAN DO IT !

SCORE	/37
mins	secs

TARGETS	/37
mins	secs

23

$186 \div 6 =$

$156 \div 12 =$ $96 \div 2 =$

$64 \div 4 =$ $90 \div 5 =$ $200 \div 8 =$

$132 \div 12 =$ $70 \div 2 =$ $98 \div 7 =$ $219 \div 3 =$

$84 \div 7 =$ $96 \div 3 =$ $176 \div 11 =$ $144 \div 4 =$ $126 \div 9 =$

$144 \div 12 =$ $126 \div 6 =$ $75 \div 5 =$ $170 \div 2 =$

$248 \div 8 =$ $168 \div 4 =$ $189 \div 9 =$ $123 \div 3 =$ $78 \div 6 =$

$86 \div 2 =$ $96 \div 12 =$ $105 \div 5 =$ $135 \div 9 =$

$155 \div 5 =$ $104 \div 4 =$ $120 \div 8 =$ $72 \div 6 =$ $154 \div 11 =$

$80 \div 5 =$ $108 \div 9 =$ $75 \div 3 =$ $147 \div 7 =$

THE DIVIDING WALL

NAME: _____

DATE: _____

24

KEY FOCUS: *Mental division of numbers up to 1000 by 2 to 12*

How many can you get right ?

How fast can you do it ?

What will you try to get next time ?

YOU CAN DO IT !

SCORE	/37
mins	secs
TARGETS	/37
mins	secs

$198 \div 9 =$

$264 \div 8 =$ $119 \div 7 =$

$430 \div 5 =$ $156 \div 4 =$ $143 \div 11 =$

$108 \div 12 =$ $138 \div 2 =$ $150 \div 6 =$ $135 \div 3 =$

$287 \div 7 =$ $488 \div 8 =$ $68 \div 4 =$ $297 \div 9 =$ $275 \div 5 =$

$253 \div 11 =$ $639 \div 9 =$ $270 \div 6 =$ $117 \div 3 =$

$92 \div 2 =$ $96 \div 8 =$ $133 \div 7 =$ $168 \div 12 =$ $65 \div 5 =$

$170 \div 5 =$ $171 \div 9 =$ $248 \div 4 =$ $121 \div 11 =$

$102 \div 6 =$ $112 \div 8 =$ $188 \div 2 =$ $161 \div 7 =$ $246 \div 3 =$

$144 \div 12 =$ $91 \div 7 =$ $185 \div 5 =$ $153 \div 9 =$

The Dividing Wall

ANSWERS

DATE	TASK	SCORE	TIME	TARGET(S)

ANSWERS

THE DIVIDING WALL

1

Total 37

What was the final score ?

How fast was it completed ?

Were the previous targets beaten ?

What are the targets for next time ?

$4 \div 2 = 2$

$6 \div 3 = 2$ $8 \div 2 = 4$

$9 \div 3 = 3$ $10 \div 5 = 2$ $12 \div 4 = 3$

$14 \div 7 = 2$ $15 \div 3 = 5$ $16 \div 4 = 4$ $18 \div 6 = 3$

$21 \div 3 = 7$ $24 \div 4 = 6$ $25 \div 5 = 5$ $27 \div 9 = 3$

$20 \div 5 = 4$ $28 \div 7 = 4$ $30 \div 5 = 6$ $32 \div 4 = 8$ $35 \div 7 = 5$

$36 \div 6 = 6$ $40 \div 5 = 8$ $42 \div 6 = 7$ $45 \div 9 = 5$ $48 \div 6 = 8$

$49 \div 7 = 7$ $50 \div 10 = 5$ $54 \div 6 = 9$ $56 \div 7 = 8$

$60 \div 10 = 6$ $63 \div 9 = 7$ $64 \div 8 = 8$ $70 \div 7 = 10$ $72 \div 8 = 9$

$80 \div 8 = 10$ $81 \div 9 = 9$ $90 \div 10 = 9$ $100 \div 10 = 10$

ANSWERS

THE DIVIDING WALL

2

Total 37

What was the final score ?

How fast was it completed ?

Were the previous targets beaten ?

What are the targets for next time ?

$100 \div 10 = 10$

$81 \div 9 = 9$ $90 \div 9 = 10$

$70 \div 7 = 10$ $72 \div 9 = 8$ $80 \div 10 = 8$

$56 \div 8 = 7$ $60 \div 10 = 6$ $63 \div 7 = 9$ $64 \div 8 = 8$ $54 \div 9 = 6$

$45 \div 9 = 5$ $48 \div 8 = 6$ $49 \div 7 = 7$ $50 \div 10 = 5$ $42 \div 7 = 6$

$35 \div 5 = 7$ $36 \div 4 = 9$ $40 \div 8 = 5$ $30 \div 5 = 6$ $32 \div 8 = 4$

$25 \div 5 = 5$ $27 \div 3 = 9$ $28 \div 7 = 4$ $21 \div 3 = 7$ $24 \div 3 = 8$

$18 \div 6 = 3$ $20 \div 4 = 5$ $14 \div 7 = 2$ $15 \div 5 = 3$ $16 \div 4 = 4$

$10 \div 2 = 5$ $12 \div 3 = 4$ $6 \div 2 = 3$ $8 \div 4 = 2$ $9 \div 3 = 3$

$4 \div 2 = 2$

ANSWERS

THE DIVIDING WALL

© 2015 Tony Colledge

Total 37

What was the final score ?

How fast was it completed ?

Were the previous targets beaten ?

What are the targets for next time ?

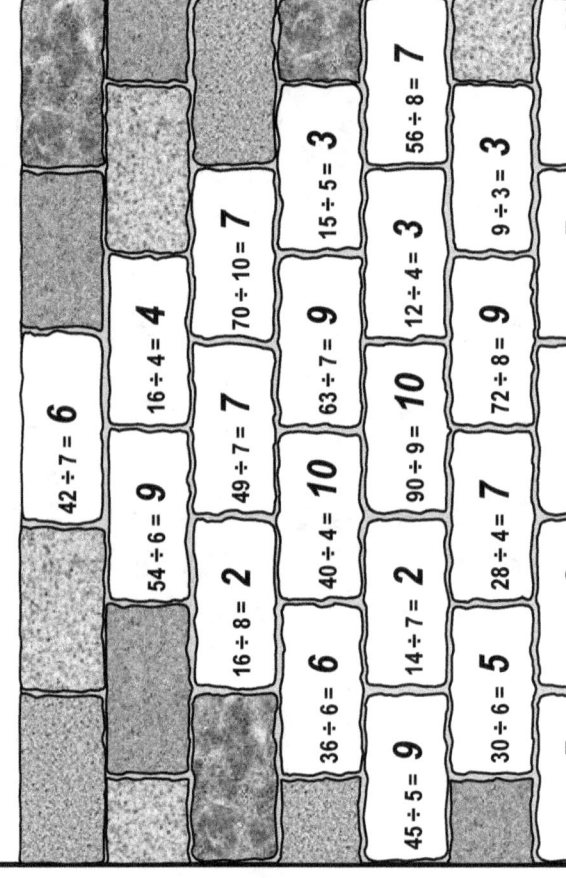

42 ÷ 7 = 6
54 ÷ 6 = 9   16 ÷ 4 = 4
16 ÷ 8 = 2   49 ÷ 7 = 7   70 ÷ 10 = 7
36 ÷ 6 = 6   40 ÷ 4 = 10   63 ÷ 7 = 9   15 ÷ 5 = 3
45 ÷ 5 = 9   14 ÷ 7 = 2   90 ÷ 9 = 10   12 ÷ 4 = 3   56 ÷ 8 = 7
25 ÷ 5 = 5   30 ÷ 6 = 5   28 ÷ 4 = 7   72 ÷ 8 = 9   9 ÷ 3 = 3
32 ÷ 8 = 4   12 ÷ 6 = 2   36 ÷ 9 = 4   40 ÷ 8 = 5   100 ÷ 10 = 10
80 ÷ 10 = 8   18 ÷ 6 = 3   50 ÷ 5 = 10   24 ÷ 4 = 6
48 ÷ 6 = 8   35 ÷ 7 = 5   81 ÷ 9 = 9
24 ÷ 8 = 3   60 ÷ 6 = 10   20 ÷ 4 = 5
18 ÷ 2 = 9   27 ÷ 3 = 9

ANSWERS

THE DIVIDING WALL

© 2015 Tony Colledge

Total 37

What was the final score ?

How fast was it completed ?

Were the previous targets beaten ?

What are the targets for next time ?

21 ÷ 3 = 7
36 ÷ 6 = 6   15 ÷ 3 = 5
14 ÷ 2 = 7   35 ÷ 7 = 5   54 ÷ 6 = 9
32 ÷ 4 = 8   72 ÷ 9 = 8   6 ÷ 2 = 3   40 ÷ 5 = 8
100 ÷ 10 = 10   27 ÷ 9 = 3   18 ÷ 2 = 9   63 ÷ 9 = 7   60 ÷ 6 = 10
56 ÷ 7 = 8   20 ÷ 5 = 4   42 ÷ 6 = 7   10 ÷ 5 = 2   24 ÷ 3 = 8
81 ÷ 9 = 9   9 ÷ 3 = 3   50 ÷ 10 = 5   64 ÷ 8 = 8   80 ÷ 8 = 10
30 ÷ 5 = 6   49 ÷ 7 = 7   4 ÷ 2 = 2   90 ÷ 10 = 9   12 ÷ 6 = 2
28 ÷ 7 = 4   45 ÷ 9 = 5   16 ÷ 2 = 8   25 ÷ 5 = 5   48 ÷ 8 = 6
8 ÷ 2 = 4   70 ÷ 7 = 10

ANSWERS

THE DIVIDING WALL

© 2015 Tony Colledge

5

Total 37

- What was the final score ?
- How fast was it completed ?
- Were the previous targets beaten ?
- What are the targets for next time ?

$56 \div 7 = 8$

$36 \div 6 = 6$ · $63 \div 9 = 7$

$40 \div 5 = 8$ · $60 \div 10 = 6$ · $64 \div 8 = 8$

$24 \div 3 = 8$ · $54 \div 9 = 6$ · $72 \div 9 = 8$ · $25 \div 5 = 5$ · $80 \div 8 = 10$

$12 \div 3 = 4$ · $40 \div 10 = 4$ · $42 \div 6 = 7$ · $18 \div 3 = 6$

$18 \div 9 = 2$ · $32 \div 4 = 8$ · $81 \div 9 = 9$ · $28 \div 7 = 4$ · $50 \div 10 = 5$

$70 \div 7 = 10$ · $16 \div 4 = 4$ · $9 \div 3 = 3$ · $45 \div 9 = 5$ · $12 \div 6 = 2$

$24 \div 6 = 4$ · $27 \div 9 = 3$ · $35 \div 5 = 7$ · $30 \div 5 = 6$ · $100 \div 10 = 10$

$48 \div 8 = 6$ · $15 \div 3 = 5$ · $36 \div 4 = 9$ · $90 \div 10 = 9$ · $16 \div 2 = 8$

$20 \div 5 = 4$ · $49 \div 7 = 7$ · $70 \div 7 = 10$

ANSWERS

THE DIVIDING WALL

© 2015 Tony Colledge

6

Total 42

- What was the final score ?
- How fast was it completed ?
- Were the previous targets beaten ?
- What are the targets for next time ?

$24 \div 4 = 6$ · $24 \div 3 = 8$

$42 \div 7 = 6$ · $10 \div 5 = 2$

$8 \div 4 = 2$ · $100 \div 10 = 10$ · $15 \div 3 = 5$

$64 \div 8 = 8$ · $16 \div 4 = 4$ · $36 \div 6 = 6$ · $21 \div 3 = 7$

$16 \div 2 = 8$ · $36 \div 4 = 9$

$45 \div 9 = 5$ · $80 \div 8 = 10$ · $27 \div 9 = 3$ · $72 \div 8 = 9$ · $14 \div 7 = 2$

$6 \div 3 = 2$ · $54 \div 6 = 9$ · $90 \div 10 = 9$ · $35 \div 7 = 5$

$32 \div 4 = 8$ · $12 \div 4 = 3$ · $40 \div 5 = 8$ · $30 \div 5 = 6$

$12 \div 6 = 2$ · $40 \div 4 = 10$

$60 \div 6 = 10$ · $81 \div 9 = 9$ · $9 \div 3 = 3$ · $56 \div 8 = 7$

$20 \div 5 = 4$ · $48 \div 6 = 8$ · $49 \div 7 = 7$ · $28 \div 7 = 4$ · $25 \div 5 = 5$

$18 \div 2 = 9$ · $70 \div 7 = 10$ · $50 \div 10 = 5$ · $18 \div 6 = 3$ · $4 \div 2 = 2$

ANSWERS
THE DIVIDING WALL

8   Total 45

What was the final score ?

How fast was it completed ?

Were the previous targets beaten ?

What are the targets for next time ?

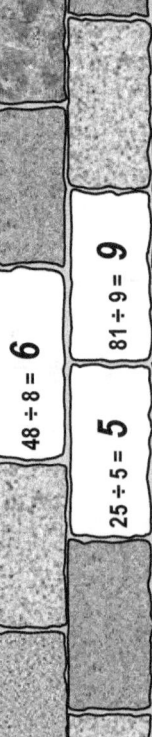

- 81 ÷ 9 = 9
- 25 ÷ 5 = 5
- 63 ÷ 7 = 9
- 72 ÷ 9 = 8
- 50 ÷ 5 = 10
- 32 ÷ 8 = 4
- 4 ÷ 2 = 2
- 20 ÷ 2 = 10
- 20 ÷ 5 = 4
- 36 ÷ 4 = 9
- 36 ÷ 6 = 6
- 90 ÷ 10 = 9
- 6 ÷ 2 = 3
- 56 ÷ 7 = 8
- 10 ÷ 2 = 5
- 12 ÷ 4 = 3
- 12 ÷ 6 = 2
- 30 ÷ 3 = 10
- 30 ÷ 5 = 6
- 48 ÷ 6 = 8
- 42 ÷ 7 = 6
- 28 ÷ 7 = 4
- 35 ÷ 5 = 7
- 27 ÷ 3 = 9
- 100 ÷ 10 = 10
- 9 ÷ 3 = 3
- 54 ÷ 9 = 6
- 14 ÷ 2 = 7
- 40 ÷ 5 = 8
- 40 ÷ 4 = 10
- 16 ÷ 2 = 8
- 16 ÷ 4 = 4
- 15 ÷ 3 = 5
- 8 ÷ 2 = 4
- 64 ÷ 8 = 8
- 70 ÷ 10 = 7
- 45 ÷ 5 = 9
- 49 ÷ 7 = 7
- 80 ÷ 8 = 10
- 18 ÷ 3 = 6
- 18 ÷ 2 = 9
- 24 ÷ 4 = 6
- 24 ÷ 3 = 8
- 21 ÷ 7 = 3

ANSWERS
THE DIVIDING WALL

7   Total 42

What was the final score ?

How fast was it completed ?

Were the previous targets beaten ?

What are the targets for next time ?

- 48 ÷ 8 = 6
- 25 ÷ 5 = 5
- 81 ÷ 9 = 9
- 100 ÷ 10 = 10
- 18 ÷ 9 = 2
- 56 ÷ 8 = 7
- 49 ÷ 7 = 7
- 30 ÷ 5 = 6
- 63 ÷ 9 = 7
- 54 ÷ 6 = 9
- 36 ÷ 6 = 6
- 36 ÷ 9 = 4
- 8 ÷ 4 = 2
- 12 ÷ 3 = 4
- 12 ÷ 2 = 6
- 70 ÷ 7 = 10
- 42 ÷ 6 = 7
- 50 ÷ 10 = 5
- 45 ÷ 9 = 5
- 21 ÷ 3 = 7
- 18 ÷ 3 = 6
- 72 ÷ 8 = 9
- 40 ÷ 8 = 5
- 40 ÷ 4 = 10
- 35 ÷ 7 = 5
- 80 ÷ 8 = 10
- 90 ÷ 10 = 9
- 32 ÷ 4 = 8
- 15 ÷ 5 = 3
- 28 ÷ 4 = 7
- 6 ÷ 3 = 2
- 24 ÷ 6 = 4
- 24 ÷ 8 = 3
- 16 ÷ 4 = 4
- 16 ÷ 8 = 2
- 9 ÷ 3 = 3
- 20 ÷ 4 = 5
- 64 ÷ 8 = 8
- 27 ÷ 9 = 3
- 60 ÷ 6 = 10

ANSWERS

THE DIVIDING WALL

9 Total **45**

What was the final score ?

How fast was it completed ?

Were the previous targets beaten ?

What are the targets for next time ?

32 ÷ 4 = 8

64 ÷ 8 = 8   49 ÷ 7 = 7

70 ÷ 7 = 10   45 ÷ 9 = 5   9 ÷ 3 = 3

54 ÷ 6 = 9   14 ÷ 7 = 2   72 ÷ 8 = 9

21 ÷ 3 = 7   28 ÷ 4 = 7   6 ÷ 3 = 2   42 ÷ 6 = 7   81 ÷ 9 = 9

30 ÷ 3 = 10   16 ÷ 4 = 4   36 ÷ 6 = 6   20 ÷ 10 = 2

30 ÷ 5 = 6   16 ÷ 8 = 2   36 ÷ 9 = 4   20 ÷ 4 = 5

27 ÷ 9 = 3   48 ÷ 8 = 6   4 ÷ 2 = 2   60 ÷ 6 = 10

80 ÷ 10 = 8   40 ÷ 8 = 5   12 ÷ 2 = 6   18 ÷ 6 = 3

24 ÷ 6 = 4   40 ÷ 10 = 4   12 ÷ 3 = 4   18 ÷ 9 = 2

50 ÷ 10 = 5   10 ÷ 5 = 2   56 ÷ 8 = 7   15 ÷ 5 = 3   100 ÷ 10 = 10

24 ÷ 8 = 3   63 ÷ 9 = 7   8 ÷ 4 = 2

90 ÷ 9 = 10   25 ÷ 5 = 5   35 ÷ 7 = 5

ANSWERS

THE DIVIDING WALL

10 Total **45**

What was the final score ?

How fast was it completed ?

Were the previous targets beaten ?

What are the targets for next time ?

63 ÷ 7 = 9

56 ÷ 7 = 8   81 ÷ 9 = 9

10 ÷ 2 = 5   64 ÷ 8 = 8   35 ÷ 5 = 7

12 ÷ 6 = 2   12 ÷ 4 = 3   30 ÷ 10 = 3   30 ÷ 6 = 5

6 ÷ 2 = 3   90 ÷ 10 = 9   28 ÷ 7 = 4   25 ÷ 5 = 5   50 ÷ 5 = 10

20 ÷ 2 = 10   20 ÷ 5 = 4   48 ÷ 6 = 8   49 ÷ 7 = 7   72 ÷ 9 = 8   18 ÷ 3 = 6   18 ÷ 2 = 9

21 ÷ 7 = 3   42 ÷ 7 = 6   14 ÷ 2 = 7   100 ÷ 10 = 10   8 ÷ 2 = 4   32 ÷ 8 = 4

36 ÷ 4 = 9   36 ÷ 6 = 6   9 ÷ 3 = 3   70 ÷ 10 = 7   24 ÷ 3 = 8   24 ÷ 4 = 6

4 ÷ 2 = 2   54 ÷ 9 = 6   15 ÷ 3 = 5   80 ÷ 8 = 10   27 ÷ 3 = 9

16 ÷ 2 = 8   16 ÷ 4 = 4   60 ÷ 10 = 6   45 ÷ 5 = 9   40 ÷ 8 = 5   40 ÷ 10 = 4

ANSWERS

THE DIVIDING WALL

© 2015 Tony Colledge

12

Total
50

What was the final score ?

How fast was it completed ?

Were the previous targets beaten ?

What are the targets for next time ?

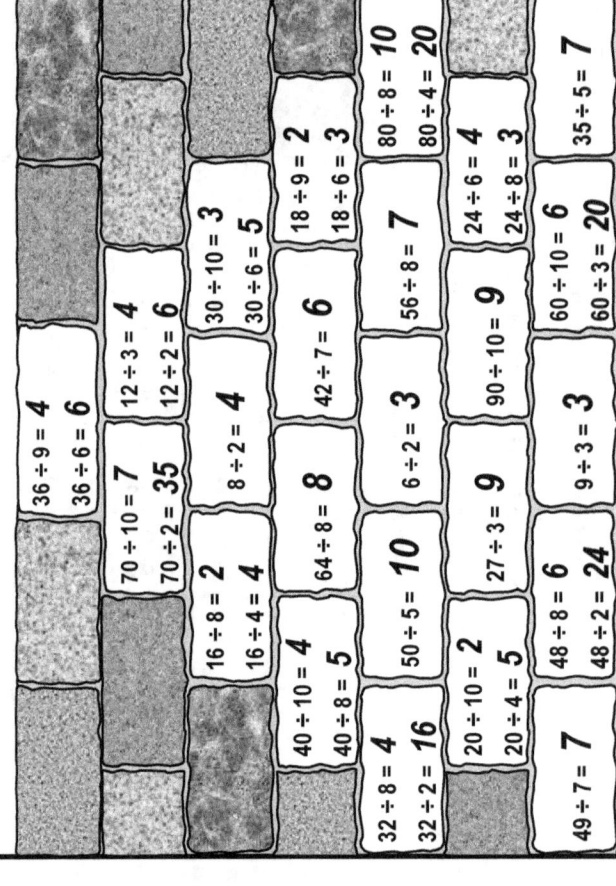

36 ÷ 9 = **4**	
36 ÷ 6 = **6**	

70 ÷ 10 = **7**   12 ÷ 3 = **4**
70 ÷ 2 = **35**   12 ÷ 2 = **6**

16 ÷ 8 = **2**   30 ÷ 10 = **3**
16 ÷ 4 = **4**   30 ÷ 6 = **5**

40 ÷ 10 = **4**   8 ÷ 2 = **4**   18 ÷ 9 = **2**
40 ÷ 8 = **5**   18 ÷ 6 = **3**

32 ÷ 8 = **4**   64 ÷ 8 = **8**   42 ÷ 7 = **6**   80 ÷ 8 = **10**
32 ÷ 2 = **16**   80 ÷ 4 = **20**

20 ÷ 10 = **2**   50 ÷ 5 = **10**   6 ÷ 2 = **3**   56 ÷ 8 = **7**
20 ÷ 4 = **5**

48 ÷ 8 = **6**   27 ÷ 3 = **9**   90 ÷ 10 = **9**   24 ÷ 6 = **4**
48 ÷ 2 = **24**   24 ÷ 8 = **3**

49 ÷ 7 = **7**   9 ÷ 3 = **3**   60 ÷ 10 = **6**   35 ÷ 5 = **7**
                      60 ÷ 3 = **20**

10 ÷ 5 = **2**   72 ÷ 9 = **8**   63 ÷ 7 = **9**   14 ÷ 2 = **7**

25 ÷ 5 = **5**   54 ÷ 9 = **6**   4 ÷ 2 = **2**   100 ÷ 10 = **10**   21 ÷ 7 = **3**

45 ÷ 9 = **5**   15 ÷ 5 = **3**   28 ÷ 4 = **7**   81 ÷ 9 = **9**

ANSWERS

THE DIVIDING WALL

© 2015 Tony Colledge

11

Total
50

What was the final score ?

How fast was it completed ?

Were the previous targets beaten ?

What are the targets for next time ?

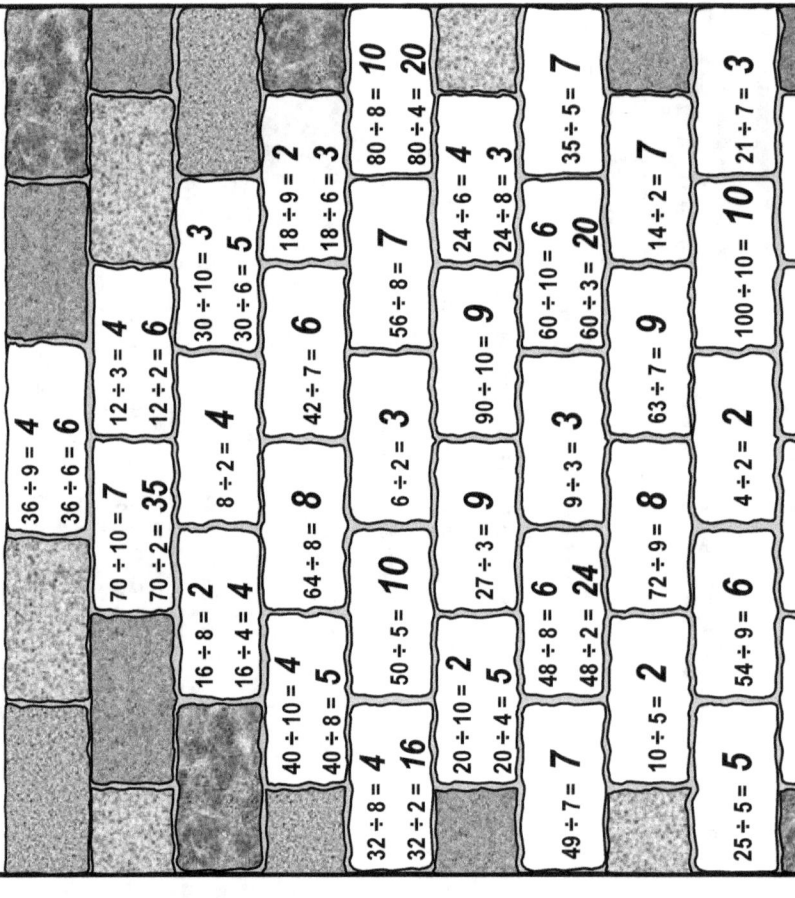

54 ÷ 6 = **9**

21 ÷ 3 = **7**   45 ÷ 9 = **5**

42 ÷ 6 = **7**   32 ÷ 4 = **8**   25 ÷ 5 = **5**
           32 ÷ 2 = **16**

35 ÷ 7 = **5**   40 ÷ 5 = **8**   18 ÷ 2 = **9**   72 ÷ 8 = **9**
           40 ÷ 4 = **10**   18 ÷ 3 = **6**

90 ÷ 9 = **10**   16 ÷ 4 = **4**   4 ÷ 2 = **2**   48 ÷ 6 = **8**   15 ÷ 3 = **5**
           16 ÷ 2 = **8**   48 ÷ 2 = **24**

           30 ÷ 5 = **6**   27 ÷ 9 = **3**   64 ÷ 8 = **8**   20 ÷ 2 = **10**   36 ÷ 6 = **6**
24 ÷ 4 = **6**   30 ÷ 3 = **10**   49 ÷ 7 = **9**   6 ÷ 3 = **2**   20 ÷ 5 = **4**   36 ÷ 4 = **9**
24 ÷ 3 = **8**   81 ÷ 9 = **9**

70 ÷ 7 = **10**   12 ÷ 4 = **3**   9 ÷ 3 = **3**   63 ÷ 9 = **7**   80 ÷ 10 = **8**   60 ÷ 10 = **6**   100 ÷ 10 = **10**
70 ÷ 2 = **35**   12 ÷ 6 = **2**   8 ÷ 2 = **4**   80 ÷ 4 = **20**   60 ÷ 3 = **20**

56 ÷ 7 = **8**   14 ÷ 2 = **7**   50 ÷ 10 = **5**   28 ÷ 7 = **4**   10 ÷ 5 = **2**

ANSWERS

THE DIVIDING WALL — 13

© 2015 Tony Colledge

Total 50

What was the final score ?

How fast was it completed ?

Were the previous targets beaten ?

What are the targets for next time ?

- 18 ÷ 2 = 9
- 18 ÷ 3 = 6
- 50 ÷ 10 = 5
- 50 ÷ 2 = 25
- 36 ÷ 6 = 6
- 36 ÷ 4 = 9
- 32 ÷ 4 = 8
- 32 ÷ 2 = 16
- 10 ÷ 5 = 2
- 72 ÷ 8 = 9
- 30 ÷ 5 = 6
- 30 ÷ 3 = 10
- 90 ÷ 9 = 10
- 90 ÷ 3 = 30
- 27 ÷ 9 = 3
- 100 ÷ 10 = 10
- 56 ÷ 7 = 8
- 12 ÷ 4 = 3
- 12 ÷ 6 = 2
- 49 ÷ 7 = 7
- 64 ÷ 8 = 8
- 16 ÷ 2 = 8
- 16 ÷ 4 = 4
- 14 ÷ 7 = 2
- 54 ÷ 6 = 9
- 9 ÷ 3 = 3
- 20 ÷ 2 = 10
- 20 ÷ 5 = 4
- 6 ÷ 3 = 2
- 63 ÷ 9 = 7
- 28 ÷ 7 = 4
- 60 ÷ 10 = 6
- 25 ÷ 5 = 5
- 24 ÷ 3 = 8
- 24 ÷ 4 = 6
- 40 ÷ 4 = 10
- 40 ÷ 5 = 8
- 4 ÷ 2 = 2
- 45 ÷ 9 = 5
- 21 ÷ 3 = 7
- 70 ÷ 7 = 10
- 8 ÷ 2 = 4
- 81 ÷ 9 = 9
- 35 ÷ 7 = 5
- 48 ÷ 6 = 8
- 48 ÷ 2 = 24
- 15 ÷ 3 = 5
- 42 ÷ 6 = 7
- 80 ÷ 10 = 8
- 80 ÷ 4 = 20

THE DIVIDING WALL — 14

© 2015 Tony Colledge

Total 50

What was the final score ?

How fast was it completed ?

Were the previous targets beaten ?

What are the targets for next time ?

- 54 ÷ 9 = 6
- 32 ÷ 8 = 4
- 21 ÷ 7 = 3
- 10 ÷ 2 = 5
- 45 ÷ 5 = 9
- 72 ÷ 9 = 8
- 36 ÷ 9 = 4
- 36 ÷ 6 = 6
- 15 ÷ 5 = 3
- 28 ÷ 7 = 4
- 28 ÷ 2 = 14
- 27 ÷ 3 = 9
- 30 ÷ 6 = 5
- 30 ÷ 2 = 15
- 64 ÷ 8 = 8
- 81 ÷ 9 = 9
- 12 ÷ 3 = 4
- 12 ÷ 2 = 6
- 25 ÷ 5 = 5
- 56 ÷ 8 = 7
- 16 ÷ 4 = 4
- 16 ÷ 8 = 2
- 40 ÷ 10 = 4
- 40 ÷ 8 = 5
- 6 ÷ 2 = 3
- 50 ÷ 10 = 5
- 4 ÷ 2 = 2
- 48 ÷ 6 = 8
- 48 ÷ 4 = 12
- 49 ÷ 7 = 7
- 90 ÷ 9 = 10
- 35 ÷ 5 = 7
- 18 ÷ 9 = 2
- 18 ÷ 6 = 3
- 100 ÷ 4 = 25
- 100 ÷ 5 = 20
- 14 ÷ 2 = 7
- 8 ÷ 4 = 2
- 60 ÷ 10 = 6
- 60 ÷ 3 = 20
- 42 ÷ 7 = 6
- 24 ÷ 6 = 4
- 24 ÷ 8 = 3
- 80 ÷ 8 = 10
- 20 ÷ 10 = 2
- 20 ÷ 4 = 5
- 9 ÷ 3 = 3
- 63 ÷ 9 = 7
- 70 ÷ 7 = 10
- 70 ÷ 2 = 35

ANSWERS

THE DIVIDING WALL

15

Total 50

What was the final score ?

How fast was it completed ?

Were the previous targets beaten ?

What are the targets for next time ?

$40 \div 5 = 8$
$40 \div 4 = 10$

$27 \div 9 = 3$ $56 \div 7 = 8$ $18 \div 3 = 6$ / $18 \div 2 = 9$ $42 \div 6 = 7$ $50 \div 10 = 5$ / $50 \div 2 = 25$

$72 \div 8 = 9$ $63 \div 7 = 9$ $32 \div 4 = 8$ $90 \div 9 = 10$ / $90 \div 2 = 45$ $81 \div 9 = 9$ $64 \div 8 = 8$ $20 \div 2 = 10$ / $20 \div 5 = 4$ $25 \div 5 = 5$

$10 \div 5 = 2$ $49 \div 7 = 7$ $6 \div 3 = 2$ $24 \div 3 = 8$ / $24 \div 4 = 6$ $9 \div 3 = 3$ $4 \div 2 = 2$ $48 \div 6 = 8$ / $48 \div 4 = 12$ $21 \div 3 = 7$

$12 \div 6 = 2$ / $12 \div 4 = 3$ $36 \div 6 = 6$ / $36 \div 9 = 4$ $70 \div 7 = 10$ $45 \div 9 = 5$ $60 \div 10 = 6$ / $60 \div 4 = 15$ $35 \div 7 = 5$ $54 \div 6 = 9$ $80 \div 10 = 8$ / $80 \div 4 = 20$

$15 \div 3 = 5$ $30 \div 5 = 6$ / $30 \div 3 = 10$ $8 \div 2 = 4$ $14 \div 7 = 2$ $2 \div 1 = 2$ $16 \div 2 = 8$ / $16 \div 4 = 4$

$100 \div 4 = 25$ $28 \div 7 = 4$

ANSWERS

THE DIVIDING WALL

16

Total 37

What was the final score ?

How fast was it completed ?

Were the previous targets beaten ?

What are the targets for next time ?

$54 \div 6 = 9$

$72 \div 9 = 8$ $36 \div 3 = 12$

$121 \div 11 = 11$ $24 \div 8 = 3$ $144 \div 12 = 12$

$120 \div 12 = 10$ $49 \div 7 = 7$ $36 \div 4 = 9$

$27 \div 9 = 3$ $48 \div 4 = 12$ $64 \div 8 = 8$ $100 \div 5 = 20$

$28 \div 4 = 7$ $21 \div 7 = 3$ $45 \div 9 = 5$

$42 \div 7 = 6$ $132 \div 12 = 11$ $60 \div 3 = 20$ $36 \div 6 = 6$

$24 \div 2 = 12$ $63 \div 7 = 9$ $99 \div 9 = 11$

$56 \div 8 = 7$ $30 \div 5 = 6$ $44 \div 4 = 11$ $108 \div 12 = 9$

$48 \div 6 = 8$ $25 \div 5 = 5$ $110 \div 11 = 10$ $18 \div 3 = 6$

$40 \div 8 = 5$ $81 \div 9 = 9$ $32 \div 8 = 4$

$24 \div 6 = 4$ $60 \div 12 = 5$

ANSWERS

THE DIVIDING WALL

18

Total 37

What was the final score ?

How fast was it completed ?

Were the previous targets beaten ?

What are the targets for next time ?

$63 \div 7 = 9$

$144 \div 12 = 12$ $40 \div 8 = 5$

$60 \div 12 = 5$ $24 \div 8 = 3$ $99 \div 11 = 9$

$36 \div 3 = 12$ $21 \div 7 = 3$ $28 \div 4 = 7$ $64 \div 8 = 8$

$56 \div 8 = 7$ $35 \div 5 = 7$ $81 \div 9 = 9$ $42 \div 7 = 6$ $132 \div 12 = 11$

$32 \div 8 = 4$ $110 \div 11 = 10$ $18 \div 3 = 6$ $27 \div 9 = 3$

$48 \div 4 = 12$ $25 \div 5 = 5$ $108 \div 12 = 9$ $36 \div 4 = 9$ $60 \div 4 = 15$

$45 \div 9 = 5$ $80 \div 4 = 20$ $24 \div 2 = 12$ $121 \div 11 = 11$

$77 \div 7 = 11$ $36 \div 6 = 6$ $54 \div 9 = 6$ $30 \div 5 = 6$ $48 \div 6 = 8$

$24 \div 6 = 4$ $120 \div 12 = 10$ $72 \div 9 = 8$ $49 \div 7 = 7$

ANSWERS

THE DIVIDING WALL

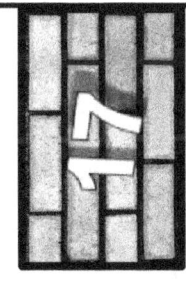

17

Total 37

What was the final score ?

How fast was it completed ?

Were the previous targets beaten ?

What are the targets for next time ?

$108 \div 9 = 12$

$33 \div 3 = 11$ $56 \div 7 = 8$

$81 \div 9 = 9$ $48 \div 8 = 6$ $24 \div 12 = 2$

$32 \div 4 = 8$ $88 \div 8 = 11$ $63 \div 9 = 7$ $30 \div 6 = 5$

$60 \div 5 = 12$ $21 \div 3 = 7$ $132 \div 11 = 12$ $24 \div 4 = 6$

$36 \div 9 = 4$ $121 \div 11 = 11$ $45 \div 5 = 9$ $80 \div 4 = 20$

$54 \div 6 = 9$ $60 \div 2 = 30$ $72 \div 8 = 9$ $36 \div 12 = 3$

$24 \div 3 = 8$ $33 \div 11 = 3$ $27 \div 3 = 9$ $40 \div 5 = 8$

$35 \div 7 = 5$ $144 \div 12 = 12$ $25 \div 5 = 5$ $28 \div 7 = 4$

$42 \div 6 = 7$ $48 \div 12 = 4$ $64 \div 8 = 8$ $110 \div 10 = 11$ $100 \div 4 = 25$

$36 \div 6 = 6$

ANSWERS
THE DIVIDING WALL

19

Total 37

© 2015 Tony Colledge

What was the final score ?

How fast was it completed ?

Were the previous targets beaten ?

What are the targets for next time ?

64 ÷ 8 = 8

56 ÷ 7 = 8   36 ÷ 9 = 4

100 ÷ 5 = 20   49 ÷ 7 = 7   90 ÷ 3 = 30

42 ÷ 6 = 7   60 ÷ 5 = 12   24 ÷ 4 = 6   72 ÷ 8 = 9

108 ÷ 9 = 12   144 ÷ 12 = 12   33 ÷ 3 = 11   63 ÷ 9 = 7   32 ÷ 4 = 8

24 ÷ 12 = 2   121 ÷ 11 = 11   70 ÷ 2 = 35

54 ÷ 6 = 9   36 ÷ 12 = 3   81 ÷ 9 = 9   16 ÷ 4 = 4   40 ÷ 5 = 8

25 ÷ 5 = 5   48 ÷ 8 = 6   21 ÷ 3 = 7   110 ÷ 10 = 11

28 ÷ 7 = 4   132 ÷ 11 = 12   18 ÷ 6 = 3   45 ÷ 5 = 9   30 ÷ 6 = 5

36 ÷ 6 = 6   24 ÷ 3 = 8   48 ÷ 12 = 4   35 ÷ 7 = 5

ANSWERS
THE DIVIDING WALL

20

Total 37

© 2015 Tony Colledge

What was the final score ?

How fast was it completed ?

Were the previous targets beaten ?

What are the targets for next time ?

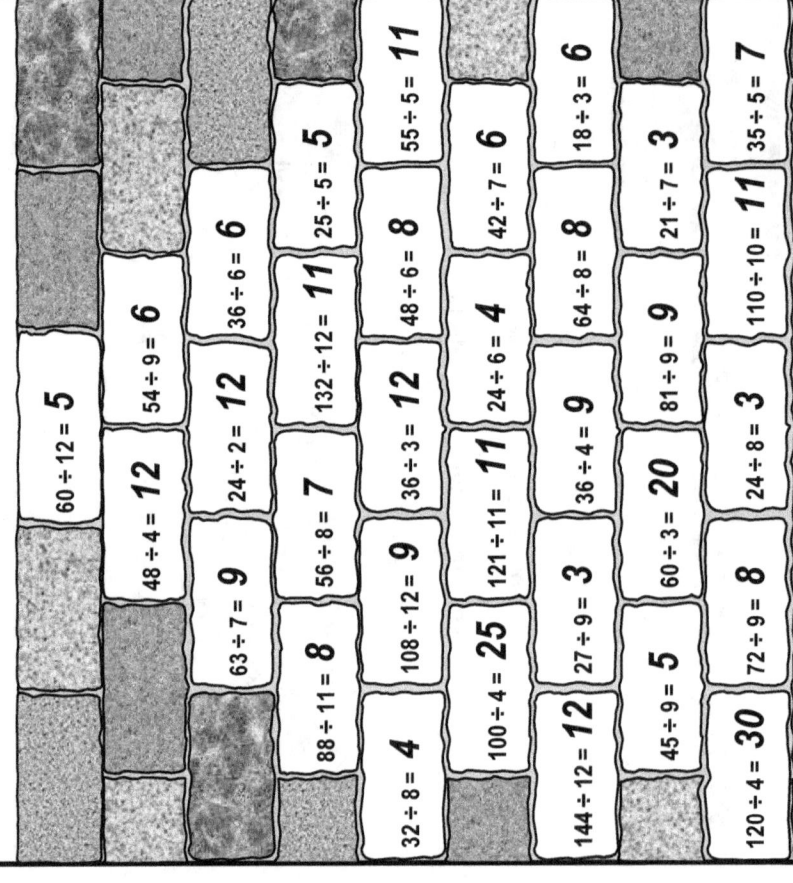

60 ÷ 12 = 5

48 ÷ 4 = 12   54 ÷ 9 = 6   24 ÷ 2 = 12   36 ÷ 6 = 6

63 ÷ 7 = 9   56 ÷ 8 = 7   36 ÷ 3 = 12   132 ÷ 12 = 11   48 ÷ 6 = 8   25 ÷ 5 = 5   55 ÷ 5 = 11

88 ÷ 11 = 8   108 ÷ 12 = 9   121 ÷ 11 = 11   24 ÷ 6 = 4   42 ÷ 7 = 6

32 ÷ 8 = 4   100 ÷ 4 = 25   27 ÷ 9 = 3   36 ÷ 4 = 9   64 ÷ 8 = 8   18 ÷ 3 = 6

144 ÷ 12 = 12   45 ÷ 9 = 5   60 ÷ 3 = 20   81 ÷ 9 = 9   21 ÷ 7 = 3

120 ÷ 4 = 30   72 ÷ 9 = 8   24 ÷ 8 = 3   110 ÷ 10 = 11   35 ÷ 5 = 7

40 ÷ 8 = 5   28 ÷ 4 = 7   49 ÷ 7 = 7   30 ÷ 5 = 6

ANSWERS

THE DIVIDING WALL — 21

Total 37

What was the final score ?

How fast was it completed ?

Were the previous targets beaten ?

What are the targets for next time ?

420 ÷ 6 = **70**

240 ÷ 12 = **20**   280 ÷ 7 = **40**

540 ÷ 9 = **60**   160 ÷ 8 = **20**   360 ÷ 6 = **60**

220 ÷ 11 = **20**   490 ÷ 7 = **70**   240 ÷ 3 = **80**   640 ÷ 8 = **80**   250 ÷ 5 = **50**

480 ÷ 6 = **80**   450 ÷ 9 = **50**   630 ÷ 7 = **90**   180 ÷ 2 = **90**   350 ÷ 7 = **50**

160 ÷ 8 = **20**   350 ÷ 5 = **70**   300 ÷ 6 = **50**   360 ÷ 9 = **40**   720 ÷ 8 = **90**

240 ÷ 6 = **40**   360 ÷ 12 = **30**   480 ÷ 8 = **60**   320 ÷ 4 = **80**   120 ÷ 3 = **40**

150 ÷ 3 = **50**   240 ÷ 8 = **30**   160 ÷ 4 = **40**   560 ÷ 7 = **80**   400 ÷ 8 = **50**

180 ÷ 3 = **60**   270 ÷ 9 = **30**   360 ÷ 4 = **90**   630 ÷ 9 = **70**   540 ÷ 6 = **90**

90 ÷ 3 = **30**

ANSWERS

THE DIVIDING WALL — 22

Total 37

What was the final score ?

How fast was it completed ?

Were the previous targets beaten ?

What are the targets for next time ?

120 ÷ 5 = **24**

168 ÷ 8 = **21**   132 ÷ 11 = **12**

108 ÷ 4 = **27**   720 ÷ 9 = **80**   150 ÷ 3 = **50**

217 ÷ 7 = **31**   110 ÷ 2 = **55**   170 ÷ 5 = **34**   426 ÷ 6 = **71**

770 ÷ 11 = **70**   240 ÷ 8 = **30**   190 ÷ 2 = **95**   180 ÷ 5 = **36**   249 ÷ 3 = **83**

90 ÷ 6 = **15**   130 ÷ 2 = **65**   819 ÷ 9 = **91**   350 ÷ 7 = **50**

180 ÷ 4 = **45**   120 ÷ 6 = **20**   186 ÷ 3 = **62**   96 ÷ 12 = **8**   116 ÷ 2 = **58**

240 ÷ 8 = **30**   98 ÷ 2 = **49**   360 ÷ 12 = **30**   450 ÷ 9 = **50**

630 ÷ 7 = **90**   140 ÷ 7 = **20**   540 ÷ 9 = **60**   120 ÷ 4 = **30**   330 ÷ 11 = **30**

105 ÷ 5 = **21**   150 ÷ 2 = **75**   216 ÷ 3 = **72**   189 ÷ 9 = **21**

ANSWERS

THE DIVIDING WALL

© 2015 Tony Colledge

24   Total 37

What was the final score ?

How fast was it completed ?

Were the previous targets beaten ?

What are the targets for next time ?

$198 \div 9 = 22$

$264 \div 8 = 33$ $119 \div 7 = 17$

$430 \div 5 = 86$ $156 \div 4 = 39$ $143 \div 11 = 13$

$108 \div 12 = 9$ $138 \div 2 = 69$ $150 \div 6 = 25$ $135 \div 3 = 45$

$287 \div 7 = 41$ $488 \div 8 = 61$ $68 \div 4 = 17$ $297 \div 9 = 33$ $275 \div 5 = 55$

$253 \div 11 = 23$ $639 \div 9 = 71$ $270 \div 6 = 45$ $117 \div 3 = 39$

$92 \div 2 = 46$ $96 \div 8 = 12$ $133 \div 7 = 19$ $168 \div 12 = 14$ $65 \div 5 = 13$

$170 \div 5 = 34$ $171 \div 9 = 19$ $248 \div 4 = 62$ $121 \div 11 = 11$

$102 \div 6 = 17$ $112 \div 8 = 14$ $188 \div 2 = 94$ $161 \div 7 = 23$ $246 \div 3 = 82$

$144 \div 12 = 12$ $91 \div 7 = 13$ $185 \div 5 = 37$ $153 \div 9 = 17$

ANSWERS

THE DIVIDING WALL

© 2015 Tony Colledge

23   Total 37

What was the final score ?

How fast was it completed ?

Were the previous targets beaten ?

What are the targets for next time ?

$186 \div 6 = 31$

$156 \div 12 = 13$ $96 \div 2 = 48$

$64 \div 4 = 16$ $90 \div 5 = 18$ $200 \div 8 = 25$

$70 \div 2 = 35$ $98 \div 7 = 14$ $219 \div 3 = 73$

$132 \div 12 = 11$ $176 \div 11 = 16$ $144 \div 4 = 36$ $126 \div 9 = 14$

$144 \div 12 = 12$ $126 \div 6 = 21$ $75 \div 5 = 15$ $170 \div 2 = 85$

$168 \div 4 = 42$ $189 \div 9 = 21$ $123 \div 3 = 41$ $78 \div 6 = 13$

$96 \div 12 = 8$ $105 \div 5 = 21$ $135 \div 9 = 15$

$248 \div 8 = 31$ $86 \div 2 = 43$ $120 \div 8 = 15$ $72 \div 6 = 12$ $154 \div 11 = 14$

$155 \div 5 = 31$ $104 \div 4 = 26$ $108 \div 9 = 12$ $75 \div 3 = 25$

$80 \div 5 = 16$ $147 \div 7 = 21$

Beat The Dividing Wall

Sheets 1 to 10 test dividing by 2 to 10 where the divisor is missing

Sheets 11 to 15 also include using inverses to find the dividend

Sheets 16 to 20 test finding factor pairs up to 10 when given the dividend

Sheets 21 to 25 test dividing by 2 to 12 where the divisor is missing

Sheets 26 to 30 test finding factor pairs up to 12 when given the dividend

Where factor pairs are required, remind pupils to only use numbers in the range indicated on the sheet. Although negative factor pairs are **mathematically correct,** and can be discussed if the situation arises, the answer sheets do not cover the option of using negative numbers.

Where answers are given as factor pairs (like 5, 8) I usually said, "5 with 8" when reading them out so the pupils understood that they can be reversed.
Where there are 2 possible answers on some bricks e.g. for 24 then different factor pairs must be used. They cannot use 4, 6 and also 6, 4.
Bricks with 2 answers get a mark for each different factor pair.

Answers are provided at the end of this section.

An example of a pupil record sheet is provided with the answers.

NAME: _____   DATE: _____

BEAT THE DIVIDING WALL

1

KEY FOCUS: *Dividing by 2 to 10*

How many can you get right ?

How fast can you do it ?

What will you try to get next time ?

YOU CAN DO IT !

SCORE	/37
mins	secs

TARGETS	/37
mins	secs

$4 \div \quad = 2$

$6 \div \quad = 3$ $8 \div \quad = 2$

$9 \div \quad = 3$ $10 \div \quad = 2$ $12 \div \quad = 6$

$14 \div \quad = 7$ $15 \div \quad = 5$ $16 \div \quad = 4$ $18 \div \quad = 2$

$20 \div \quad = 10$ $21 \div \quad = 3$ $24 \div \quad = 6$ $25 \div \quad = 5$ $27 \div \quad = 3$

$28 \div \quad = 7$ $30 \div \quad = 6$ $32 \div \quad = 4$ $35 \div \quad = 5$

$36 \div \quad = 6$ $40 \div \quad = 4$ $42 \div \quad = 6$ $45 \div \quad = 9$ $48 \div \quad = 8$

$49 \div \quad = 7$ $50 \div \quad = 10$ $54 \div \quad = 6$ $56 \div \quad = 7$

$60 \div \quad = 6$ $63 \div \quad = 9$ $64 \div \quad = 8$ $70 \div \quad = 10$ $72 \div \quad = 8$

$80 \div \quad = 8$ $81 \div \quad = 9$ $90 \div \quad = 10$ $100 \div \quad = 10$

NAME: _____ DATE: _____

BEAT THE DIVIDING WALL

2

KEY FOCUS: *Dividing by 2 to 10*

How many can you get right ?

How fast can you do it ?

What will you try to get next time ?

YOU CAN DO IT !

SCORE	/37
mins	secs
TARGETS	/37
mins	secs

$100 \div = 10$

$90 \div = 9$ $81 \div = 9$

$80 \div = 10$ $72 \div = 9$ $70 \div = 7$

$64 \div = 8$ $63 \div = 7$ $60 \div = 10$ $56 \div = 8$

$54 \div = 9$ $50 \div = 5$ $49 \div = 7$ $48 \div = 6$ $45 \div = 5$

$42 \div = 7$ $40 \div = 5$ $36 \div = 4$ $35 \div = 7$

$32 \div = 8$ $30 \div = 10$ $28 \div = 4$ $27 \div = 9$ $25 \div = 5$

$24 \div = 8$ $21 \div = 7$ $20 \div = 4$ $18 \div = 3$

$16 \div = 2$ $15 \div = 3$ $14 \div = 2$ $12 \div = 4$ $10 \div = 5$

$9 \div = 3$ $8 \div = 4$ $6 \div = 2$ $4 \div = 2$

NAME: _____   DATE: _____

BEAT THE DIVIDING WALL

3

KEY FOCUS: *Dividing by 2 to 10*

How many can you get right ?

How fast can you do it ?

What will you try to get next time ?

YOU CAN DO IT !

SCORE	/37
mins	secs
TARGETS	/37
mins	secs

$63 \div \quad = 9$

$64 \div \quad = 8$ $42 \div \quad = 6$

$16 \div \quad = 4$ $60 \div \quad = 6$ $9 \div \quad = 3$

$45 \div \quad = 9$ $24 \div \quad = 4$ $72 \div \quad = 8$ $20 \div \quad = 2$

$90 \div \quad = 10$ $4 \div \quad = 2$ $36 \div \quad = 6$ $25 \div \quad = 5$ $56 \div \quad = 7$

$27 \div \quad = 3$ $100 \div \quad = 10$ $12 \div \quad = 2$ $49 \div \quad = 7$

$10 \div \quad = 2$ $54 \div \quad = 6$ $30 \div \quad = 5$ $70 \div \quad = 10$ $18 \div \quad = 9$

$32 \div \quad = 4$ $15 \div \quad = 5$ $48 \div \quad = 8$ $6 \div \quad = 3$

$50 \div \quad = 10$ $81 \div \quad = 9$ $14 \div \quad = 7$ $80 \div \quad = 8$ $28 \div \quad = 7$

$8 \div \quad = 2$ $35 \div \quad = 5$ $40 \div \quad = 10$ $21 \div \quad = 3$

NAME: _____   DATE: _____

BEAT THE
DIVIDING WALL

4

KEY FOCUS: *Dividing by 2 to 10*

How many can you get right ?

How fast can you do it ?

What will you try to get next time ?

YOU CAN DO IT !

SCORE	/37
mins	**secs**
TARGETS	/37
mins	**secs**

$54 \div \quad = 9$

$28 \div \quad = 4$ | $45 \div \quad = 5$

$35 \div \quad = 7$ | $56 \div \quad = 8$ | $12 \div \quad = 3$

$72 \div \quad = 9$ | $18 \div \quad = 6$ | $40 \div \quad = 8$ | $27 \div \quad = 9$

$30 \div \quad = 3$ | $42 \div \quad = 7$ | $4 \div \quad = 2$ | $100 \div \quad = 10$ | $21 \div \quad = 7$

$60 \div \quad = 10$ | $16 \div \quad = 8$ | $49 \div \quad = 7$ | $9 \div \quad = 3$

$64 \div \quad = 8$ | $10 \div \quad = 5$ | $48 \div \quad = 6$ | $25 \div \quad = 5$ | $80 \div \quad = 10$

$6 \div \quad = 2$ | $50 \div \quad = 5$ | $14 \div \quad = 2$ | $32 \div \quad = 8$

$36 \div \quad = 9$ | $15 \div \quad = 3$ | $90 \div \quad = 9$ | $20 \div \quad = 5$ | $63 \div \quad = 7$

$24 \div \quad = 3$ | $70 \div \quad = 7$ | $8 \div \quad = 4$ | $81 \div \quad = 9$

NAME: _____   DATE: _____

BEAT THE DIVIDING WALL

KEY FOCUS: *Dividing by 2 to 10*

5

How many can you get right ?

How fast can you do it ?

What will you try to get next time ?

YOU CAN DO IT !

SCORE	/37
mins	secs

TARGETS	/37
mins	secs

81 ÷ ___ = 9

63 ÷ ___ = 9   48 ÷ ___ = 8

100 ÷ ___ = 10   72 ÷ ___ = 8   15 ÷ ___ = 5

21 ÷ ___ = 3   54 ÷ ___ = 6   28 ÷ ___ = 7   45 ÷ ___ = 9

16 ÷ ___ = 4   42 ÷ ___ = 6   64 ÷ ___ = 8   27 ÷ ___ = 3   80 ÷ ___ = 8

8 ÷ ___ = 2   70 ÷ ___ = 10   9 ÷ ___ = 3   36 ÷ ___ = 4

32 ÷ ___ = 4   20 ÷ ___ = 10   56 ÷ ___ = 7   6 ÷ ___ = 3   40 ÷ ___ = 4

35 ÷ ___ = 5   4 ÷ ___ = 2   90 ÷ ___ = 10   14 ÷ ___ = 7

10 ÷ ___ = 2   49 ÷ ___ = 7   24 ÷ ___ = 6   50 ÷ ___ = 10   12 ÷ ___ = 6

60 ÷ ___ = 6   25 ÷ ___ = 5   18 ÷ ___ = 2   30 ÷ ___ = 6

NAME: _____   DATE: _____

BEAT THE DIVIDING WALL

KEY FOCUS: *Dividing by 2 to 10*

6

How many can you get right ?

How fast can you do it ?

What will you try to get next time ?

YOU CAN DO IT !

SCORE	/45
mins	secs

TARGETS	/45
mins	secs

$72 \div = 9$

$32 \div = 8$ $49 \div = 7$

$10 \div = 5$ $64 \div = 8$ $90 \div = 9$

$12 \div = 6$
$12 \div = 4$ $56 \div = 8$ $27 \div = 9$ $20 \div = 4$
$20 \div = 10$

$60 \div = 10$ $15 \div = 3$ $81 \div = 9$ $6 \div = 2$ $48 \div = 6$

$18 \div = 2$
$18 \div = 3$ $14 \div = 2$ $54 \div = 9$ $24 \div = 8$
$24 \div = 6$

$35 \div = 7$ $80 \div = 10$ $28 \div = 4$ $63 \div = 7$ $8 \div = 4$

$36 \div = 4$
$36 \div = 6$ $4 \div = 2$ $9 \div = 3$ $30 \div = 10$
$30 \div = 6$

$25 \div = 5$ $70 \div = 7$ $45 \div = 5$ $21 \div = 7$ $50 \div = 5$

$40 \div = 4$
$40 \div = 5$ $42 \div = 7$ $100 \div = 10$ $16 \div = 4$
$16 \div = 2$

NAME: _____   DATE: _____

BEAT THE DIVIDING WALL

KEY FOCUS: *Dividing by 2 to 10*

7

How many can you get right ?

How fast can you do it ?

What will you try to get next time ?

YOU CAN DO IT !

SCORE	/45
mins	secs
TARGETS	/45
mins	secs

$56 \div \quad = 8$

$45 \div \quad = 5$ $\quad$ $90 \div \quad = 9$

$40 \div \quad = 10$
$40 \div \quad = 8$ $\quad$ $63 \div \quad = 7$ $\quad$ $24 \div \quad = 3$
$24 \div \quad = 4$

$48 \div \quad = 6$ $\quad$ $27 \div \quad = 9$ $\quad$ $15 \div \quad = 3$ $\quad$ $49 \div \quad = 7$

$42 \div \quad = 7$ $\quad$ $36 \div \quad = 9$
$36 \div \quad = 6$ $\quad$ $9 \div \quad = 3$ $\quad$ $16 \div \quad = 4$
$16 \div \quad = 8$ $\quad$ $72 \div \quad = 9$

$80 \div \quad = 10$ $\quad$ $14 \div \quad = 2$ $\quad$ $50 \div \quad = 5$ $\quad$ $35 \div \quad = 7$

$25 \div \quad = 5$ $\quad$ $12 \div \quad = 3$
$12 \div \quad = 2$ $\quad$ $64 \div \quad = 8$ $\quad$ $20 \div \quad = 2$
$20 \div \quad = 5$ $\quad$ $100 \div \quad = 10$

$6 \div \quad = 2$ $\quad$ $28 \div \quad = 4$ $\quad$ $70 \div \quad = 7$ $\quad$ $8 \div \quad = 4$

$81 \div \quad = 9$ $\quad$ $30 \div \quad = 3$
$30 \div \quad = 5$ $\quad$ $4 \div \quad = 2$ $\quad$ $18 \div \quad = 6$
$18 \div \quad = 9$ $\quad$ $60 \div \quad = 10$

$21 \div \quad = 7$ $\quad$ $54 \div \quad = 9$ $\quad$ $32 \div \quad = 8$ $\quad$ $10 \div \quad = 5$

NAME: _____ DATE: _____

BEAT THE DIVIDING WALL

8

© 2015 Tony Colledge

<u>KEY FOCUS</u>: *Dividing by 2 to 10*

How many can you get right ?

How fast can you do it ?

What will you try to get next time ?

YOU CAN DO IT !

SCORE	/45
mins	secs
TARGETS	/45
mins	secs

$49 \div = 7$

$81 \div = 9$ $32 \div = 4$

$48 \div = 8$ $100 \div = 10$ $54 \div = 6$

$63 \div = 9$ $15 \div = 5$ $72 \div = 8$ $9 \div = 3$

$90 \div = 10$ $18 \div = 2$ / $18 \div = 3$ $64 \div = 8$ $40 \div = 5$ / $40 \div = 4$ $14 \div = 7$

$24 \div = 6$ / $24 \div = 8$ $35 \div = 5$ $4 \div = 2$ $30 \div = 10$ / $30 \div = 6$

$60 \div = 6$ $16 \div = 2$ / $16 \div = 4$ $6 \div = 3$ $36 \div = 9$ / $36 \div = 6$ $28 \div = 7$

$20 \div = 4$ / $20 \div = 10$ $25 \div = 5$ $70 \div = 10$ $12 \div = 4$ / $12 \div = 6$

$27 \div = 3$ $56 \div = 7$ $8 \div = 2$ $50 \div = 10$ $42 \div = 6$

$45 \div = 9$ $21 \div = 3$ $80 \div = 8$ $10 \div = 2$

NAME: _____   DATE: _____

BEAT THE DIVIDING WALL

© 2015 Tony Colledge

9

KEY FOCUS: *Dividing by 2 to 10*

How many can you get right ?

How fast can you do it ?

What will you try to get next time ?

YOU CAN DO IT !

SCORE	/45
mins	secs
TARGETS	/45
mins	secs

$48 \div \quad = 6$

$54 \div \quad = 9$ $21 \div \quad = 7$

$32 \div \quad = 8$ $63 \div \quad = 7$ $28 \div \quad = 4$

$20 \div \quad = 5$
$20 \div \quad = 2$ $70 \div \quad = 7$ $81 \div \quad = 9$ $24 \div \quad = 3$
$24 \div \quad = 4$

$8 \div \quad = 4$ $80 \div \quad = 10$ $36 \div \quad = 6$
$36 \div \quad = 9$ $56 \div \quad = 8$ $35 \div \quad = 7$

$16 \div \quad = 8$
$16 \div \quad = 4$ $45 \div \quad = 5$ $9 \div \quad = 3$ $40 \div \quad = 10$
$40 \div \quad = 8$

$42 \div \quad = 7$ $25 \div \quad = 5$ $72 \div \quad = 9$ $60 \div \quad = 10$ $6 \div \quad = 2$

$100 \div \quad = 10$ $15 \div \quad = 3$ $64 \div \quad = 8$ $27 \div \quad = 9$

$12 \div \quad = 3$
$12 \div \quad = 2$ $4 \div \quad = 2$ $30 \div \quad = 3$
$30 \div \quad = 5$ $50 \div \quad = 5$ $18 \div \quad = 6$
$18 \div \quad = 9$

$14 \div \quad = 2$ $90 \div \quad = 9$ $10 \div \quad = 5$ $49 \div \quad = 7$

NAME: _____   DATE: _____

BEAT THE DIVIDING WALL

10

KEY FOCUS: *Dividing by 2 to 10*

How many can you get right ?

How fast can you do it ?

What will you try to get next time ?

YOU CAN DO IT !

SCORE	/45
mins	secs
TARGETS	/45
mins	secs

$42 \div = 6$

$64 \div = 8$ $27 \div = 3$

$15 \div = 5$ $80 \div = 8$ $56 \div = 7$

$20 \div = 4$
$20 \div = 10$ $54 \div = 6$ $28 \div = 7$ $18 \div = 2$
$18 \div = 3$

$50 \div = 10$ $4 \div = 2$ $45 \div = 9$ $14 \div = 7$ $72 \div = 8$

$63 \div = 9$ $12 \div = 4$
$12 \div = 6$ $30 \div = 6$
$30 \div = 10$ $100 \div = 10$

$49 \div = 7$ $10 \div = 2$ $32 \div = 4$ $6 \div = 3$ $70 \div = 10$

$35 \div = 5$ $36 \div = 6$
$36 \div = 4$ $16 \div = 2$
$16 \div = 4$ $81 \div = 9$

$90 \div = 10$ $21 \div = 3$ $48 \div = 8$ $25 \div = 5$ $9 \div = 3$

$24 \div = 8$
$24 \div = 6$ $8 \div = 2$ $60 \div = 6$ $40 \div = 5$
$40 \div = 4$

NAME: _____   **DATE:** _____

BEAT THE DIVIDING WALL

KEY FOCUS: *Dividing by 2 to 10 and using inverses*

11

How many can you get right ?

How fast can you do it ?

What will you try to get next time ?

YOU CAN DO IT !

SCORE	/37
mins	secs
TARGETS	/37
mins	secs

$64 \div = 8$

$\div 9 = 3$ $49 \div = 7$

$\div 7 = 9$ $25 \div = 5$ $\div 6 = 8$

$\div 4 = 7$ $6 \div = 3$ $\div 3 = 8$ $45 \div = 9$

$\div 2 = 5$ $32 \div = 4$ $\div 9 = 10$ $21 \div = 3$ $\div 8 = 5$

$100 \div = 10$ $12 \div = 2$ $\div 7 = 10$ $56 \div = 7$

$30 \div = 6$ $\div 6 = 3$ $\div 2 = 2$ $20 \div = 10$ $\div 6 = 6$

$42 \div = 6$ $\div 7 = 5$ $8 \div = 2$ $\div 10 = 6$

$\div 3 = 3$ $14 \div = 7$ $\div 9 = 8$ $54 \div = 6$ $\div 5 = 3$

$80 \div = 8$ $\div 9 = 9$ $16 \div = 4$ $\div 5 = 10$

NAME: _____ DATE: _____

BEAT THE DIVIDING WALL

12

KEY FOCUS: *Dividing by 2 to 10 and using inverses*

How many can you get right ?

How fast can you do it ?

What will you try to get next time ?

YOU CAN DO IT !

SCORE	/37
mins	secs

TARGETS	/37
mins	secs

$63 \div = 7$

$ \div 8 = 2$ $36 \div = 9$

$72 \div = 9$ $ \div 7 = 6$ $18 \div = 2$

$ \div 9 = 6$ $48 \div = 6$ $ \div 5 = 5$ $40 \div = 10$

$81 \div = 9$ $ \div 2 = 7$ $27 \div = 9$ $ \div 8 = 8$ $70 \div = 7$

$50 \div = 5$ $ \div 8 = 7$ $15 \div = 3$ $ \div 4 = 2$

$ \div 4 = 8$ $9 \div = 3$ $ \div 5 = 4$ $60 \div = 10$ $ \div 10 = 10$

$ \div 2 = 3$ $28 \div = 4$ $ \div 5 = 9$ $10 \div = 5$

$ \div 3 = 10$ $24 \div = 6$ $ \div 10 = 8$ $35 \div = 7$ $ \div 7 = 3$

$90 \div = 9$ $ \div 7 = 7$ $4 \div = 2$ $ \div 3 = 4$

NAME: _____   DATE: _____

BEAT THE DIVIDING WALL

13

KEY FOCUS: *Dividing by 2 to 10 and using inverses*

How many can you get right ?

How fast can you do it ?

What will you try to get next time ?

YOU CAN DO IT !

SCORE	/37
mins	secs
TARGETS	/37
mins	secs

$\div 9 = 9$

$\div 8 = 5$ $56 \div \quad = 8$

$49 \div \quad = 7$ $\div 9 = 3$ $42 \div \quad = 7$

$\div 10 = 6$ $54 \div \quad = 9$ $\div 6 = 8$ $20 \div \quad = 5$

$25 \div \quad = 5$ $\div 7 = 10$ $6 \div \quad = 2$ $64 \div \quad = 8$ $\div 7 = 4$

$\div 9 = 8$ $45 \div \quad = 5$ $\div 5 = 10$ $8 \div \quad = 4$

$\div 3 = 3$ $21 \div \quad = 3$ $\div 10 = 9$ $32 \div \quad = 4$ $\div 2 = 5$

$100 \div \quad = 10$ $\div 3 = 8$ $14 \div \quad = 7$ $\div 5 = 3$

$\div 5 = 7$ $16 \div \quad = 2$ $\div 9 = 7$ $30 \div \quad = 3$ $\div 2 = 2$

$\div 6 = 6$ $80 \div \quad = 10$ $\div 2 = 9$ $12 \div \quad = 3$

NAME: _____   DATE: _____

BEAT THE
DIVIDING WALL

14

KEY FOCUS: *Dividing by 2 to 10 and using inverses*

How many can you get right ?

How fast can you do it ?

What will you try to get next time ?

YOU CAN DO IT !

SCORE	/37
mins	secs
TARGETS	/37
mins	secs

÷ 6 = 9

÷ 8 = 10   28 ÷   = 7

81 ÷   = 9   ÷ 5 = 6   63 ÷   = 9

÷ 7 = 6   24 ÷   = 6   ÷ 4 = 3   40 ÷   = 5

÷ 2 = 3   60 ÷   = 6   ÷ 9 = 5   27 ÷   = 3   ÷ 4 = 4

18 ÷   = 2   ÷ 5 = 5   48 ÷   = 8   ÷ 8 = 2

50 ÷   = 10   72 ÷   = 8   90 ÷   = 9   4 ÷   = 2   ÷ 7 = 3

÷ 10 = 10   15 ÷   = 5   ÷ 7 = 8   70 ÷   = 10

÷ 4 = 5   35 ÷   = 5   ÷ 8 = 4   10 ÷   = 2   ÷ 7 = 7

9 ÷   = 3   ÷ 7 = 2   36 ÷   = 4   ÷ 8 = 8

NAME: _____   DATE: _____

BEAT THE DIVIDING WALL

15

KEY FOCUS: *Dividing by 2 to 10 and using inverses*

How many can you get right ?

How fast can you do it ?

What will you try to get next time ?

YOU CAN DO IT !

SCORE	/37
mins	secs
TARGETS	/37
mins	secs

$64 \div = 8$

$\div 9 = 3$ $\qquad$ $45 \div = 9$

$\div 8 = 9$ $\qquad$ $56 \div = 7$ $\qquad$ $\div 10 = 10$

$\div 6 = 7$ $\quad$ $32 \div = 8$ $\qquad$ $\div 9 = 4$ $\quad$ $25 \div = 5$

$9 \div = 3$ $\qquad$ $\div 5 = 4$ $\quad$ $24 \div = 3$ $\quad$ $60 \div = 6$ $\qquad$ $\div 7 = 9$

$\div 10 = 8$ $\quad$ $12 \div = 2$ $\qquad$ $\div 7 = 7$ $\quad$ $30 \div = 6$

$\div 5 = 10$ $\quad$ $40 \div = 8$ $\qquad$ $\div 9 = 9$ $\quad$ $90 \div = 10$ $\qquad$ $\div 3 = 5$

$\div 2 = 7$ $\quad$ $6 \div = 3$ $\qquad$ $\div 6 = 9$ $\quad$ $28 \div = 4$

$\div 5 = 7$ $\quad$ $16 \div = 4$ $\qquad$ $\div 5 = 2$ $\quad$ $18 \div = 6$ $\qquad$ $\div 4 = 2$

$\div 8 = 6$ $\quad$ $4 \div = 2$ $\qquad$ $\div 3 = 7$ $\quad$ $70 \div = 10$

NAME: _____   DATE: _____

BEAT THE DIVIDING WALL

KEY FOCUS: *Dividing by 2 to 10 using factor pairs*

How many can you get right ?

How fast can you do it ?

What will you try to get next time ?

YOU CAN DO IT !

16

SCORE	/45
mins	secs
TARGETS	/45
mins	secs

$54 \div \quad =$

$72 \div \quad = \qquad 48 \div \quad =$

$100 \div \quad = \qquad 64 \div \quad = \qquad 35 \div \quad =$

$36 \div \quad =$
$36 \div \quad =$ $15 \div \quad =$ $56 \div \quad =$ $30 \div \quad =$
$30 \div \quad =$

$28 \div \quad = \qquad 25 \div \quad = \qquad 90 \div \quad = \qquad 4 \div \quad = \qquad 49 \div \quad =$

$12 \div \quad =$
$12 \div \quad =$ $50 \div \quad =$ $27 \div \quad =$ $20 \div \quad =$
$20 \div \quad =$

$9 \div \quad = \qquad 70 \div \quad = \qquad 32 \div \quad = \qquad 45 \div \quad = \qquad 21 \div \quad =$

$16 \div \quad =$
$16 \div \quad =$ $81 \div \quad =$ $60 \div \quad =$ $24 \div \quad =$
$24 \div \quad =$

$8 \div \quad = \qquad 42 \div \quad = \qquad 14 \div \quad = \qquad 80 \div \quad = \qquad 6 \div \quad =$

$40 \div \quad =$
$40 \div \quad =$ $63 \div \quad =$ $10 \div \quad =$ $18 \div \quad =$
$18 \div \quad =$

NAME: _____   DATE: _____

BEAT THE DIVIDING WALL

17

© 2015 Tony Colledge

KEY FOCUS: *Dividing by 2 to 10 using factor pairs*

How many can you get right ?

How fast can you do it ?

What will you try to get next time ?

YOU CAN DO IT !

SCORE	/45
mins	secs
TARGETS	/45
mins	secs

28 ÷   =

54 ÷   =   64 ÷   =

35 ÷   =   48 ÷   =   9 ÷   =

16 ÷   =   63 ÷   =   32 ÷   =   18 ÷   =
16 ÷   =   18 ÷   =

90 ÷   =   8 ÷   =   14 ÷   =   60 ÷   =   42 ÷   =

27 ÷   =   40 ÷   =   12 ÷   =   10 ÷   =
                     40 ÷   =   12 ÷   =

25 ÷   =   80 ÷   =   49 ÷   =   6 ÷   =   70 ÷   =

56 ÷   =   36 ÷   =   20 ÷   =   81 ÷   =
                     36 ÷   =   20 ÷   =

4 ÷   =   50 ÷   =   21 ÷   =   72 ÷   =   45 ÷   =

30 ÷   =   100 ÷   =   15 ÷   =   24 ÷   =
30 ÷   =   24 ÷   =

NAME: _____ DATE: _____

BEAT THE DIVIDING WALL

KEY FOCUS: *Dividing by 2 to 10 using factor pairs*

18

How many can you get right ?

How fast can you do it ?

What will you try to get next time ?

YOU CAN DO IT !

SCORE	/45
mins	secs
TARGETS	/45
mins	secs

$81 \div \quad =$

$45 \div \quad =$ $56 \div \quad =$

$49 \div \quad =$ $70 \div \quad =$ $28 \div \quad =$

$10 \div \quad =$ $54 \div \quad =$ $8 \div \quad =$ $100 \div \quad =$

$14 \div \quad =$ $20 \div \quad =$ $32 \div \quad =$ $18 \div \quad =$ $4 \div \quad =$
$\quad\quad\quad 20 \div \quad = \quad\quad\quad\quad\quad 18 \div \quad =$

$9 \div \quad =$ $12 \div \quad =$ $30 \div \quad =$ $42 \div \quad =$
$\quad\quad\quad 12 \div \quad = \quad 30 \div \quad =$

$90 \div \quad =$ $15 \div \quad =$ $64 \div \quad =$ $27 \div \quad =$ $80 \div \quad =$

$35 \div \quad =$ $16 \div \quad =$ $36 \div \quad =$ $50 \div \quad =$
$\quad\quad\quad 16 \div \quad = \quad 36 \div \quad =$

$21 \div \quad =$ $24 \div \quad =$ $72 \div \quad =$ $40 \div \quad =$ $6 \div \quad =$
$\quad\quad\quad 24 \div \quad = \quad\quad\quad\quad\quad 40 \div \quad =$

$60 \div \quad =$ $48 \div \quad =$ $25 \div \quad =$ $63 \div \quad =$

NAME: _____   DATE: _____

BEAT THE DIVIDING WALL

19

© 2015 Tony Colledge

KEY FOCUS: *Dividing by 2 to 10 using factor pairs*

How many can you get right ?

How fast can you do it ?

What will you try to get next time ?

YOU CAN DO IT !

SCORE	/45
mins	secs
TARGETS	/45
mins	secs

27 ÷ =

21 ÷ =   48 ÷ =

42 ÷ =   80 ÷ =   63 ÷ =

100 ÷ =   54 ÷ =   28 ÷ =   9 ÷ =

72 ÷ =   40 ÷ =   45 ÷ =   12 ÷ =   60 ÷ =
            40 ÷ =   12 ÷ =

36 ÷ =   56 ÷ =   15 ÷ =   16 ÷ =
36 ÷ =   16 ÷ =

25 ÷ =   90 ÷ =   32 ÷ =   81 ÷ =   4 ÷ =

30 ÷ =   6 ÷ =   50 ÷ =   18 ÷ =
30 ÷ =   18 ÷ =

70 ÷ =   24 ÷ =   64 ÷ =   20 ÷ =   14 ÷ =
            24 ÷ =   20 ÷ =

49 ÷ =   10 ÷ =   35 ÷ =   8 ÷ =

NAME: _____ DATE: _____

BEAT THE DIVIDING WALL

20

© 2015 Tony Colledge

<u>KEY FOCUS</u>: *Dividing by 2 to 10 using factor pairs*

How many can you get right ?

How fast can you do it ?

What will you try to get next time ?

YOU CAN DO IT !

SCORE	/45
mins	secs

TARGETS	/45
mins	secs

56 ÷   =

64 ÷   =   100 ÷   =

28 ÷   =   81 ÷   =   15 ÷   =

45 ÷   =   60 ÷   =   25 ÷   =   49 ÷   =

14 ÷   =   32 ÷   =   70 ÷   =   63 ÷   =   6 ÷   =

16 ÷   =   36 ÷   =   40 ÷   =   24 ÷   =
16 ÷   =   36 ÷   =   40 ÷   =   24 ÷   =

42 ÷   =   80 ÷   =   72 ÷   =   8 ÷   =   50 ÷   =

20 ÷   =   12 ÷   =   30 ÷   =   18 ÷   =
20 ÷   =   12 ÷   =   30 ÷   =   18 ÷   =

27 ÷   =   10 ÷   =   35 ÷   =   9 ÷   =   54 ÷   =

21 ÷   =   4 ÷   =   90 ÷   =   48 ÷   =

NAME: _____   DATE: _____

BEAT THE DIVIDING WALL

© 2015 Tony Colledge   **KEY FOCUS**: *Dividing by 2 to 12*

21

How many can you get right ?

How fast can you do it ?

What will you try to get next time ?

YOU CAN DO IT !

SCORE	/37
mins	secs

TARGETS	/37
mins	secs

144 ÷   = 12

121 ÷   = 11   132 ÷   = 12

96 ÷   = 8   108 ÷   = 12   120 ÷   = 10

72 ÷   = 6   77 ÷   = 7   81 ÷   = 9   84 ÷   = 12

60 ÷   = 12   60 ÷   = 10   63 ÷   = 7   64 ÷   = 8   72 ÷   = 9

49 ÷   = 7   54 ÷   = 6   55 ÷   = 11   56 ÷   = 8

42 ÷   = 7   44 ÷   = 11   45 ÷   = 5   48 ÷   = 6   48 ÷   = 4

36 ÷   = 6   36 ÷   = 9   36 ÷   = 12   40 ÷   = 8

25 ÷   = 5   27 ÷   = 3   28 ÷   = 4   32 ÷   = 8   35 ÷   = 5

21 ÷   = 7   24 ÷   = 6   24 ÷   = 3   24 ÷   = 2

NAME: _____   DATE: _____

BEAT THE DIVIDING WALL

22

KEY FOCUS: *Dividing by 2 to 12*

How many can you get right ?

How fast can you do it ?

What will you try to get next time ?

YOU CAN DO IT !

SCORE	/37
mins	secs
TARGETS	/37
mins	secs

$121 \div \quad = 11$

$63 \div \quad = 9$ $32 \div \quad = 4$

$45 \div \quad = 9$ $96 \div \quad = 12$ $24 \div \quad = 8$

$144 \div \quad = 12$ $21 \div \quad = 3$ $54 \div \quad = 9$ $36 \div \quad = 4$

$36 \div \quad = 3$ $56 \div \quad = 7$ $42 \div \quad = 6$ $77 \div \quad = 11$ $48 \div \quad = 8$

$44 \div \quad = 4$ $72 \div \quad = 8$ $27 \div \quad = 9$ $132 \div \quad = 12$

$49 \div \quad = 7$ $108 \div \quad = 9$ $28 \div \quad = 7$ $60 \div \quad = 5$ $25 \div \quad = 5$

$24 \div \quad = 12$ $64 \div \quad = 8$ $35 \div \quad = 7$ $72 \div \quad = 12$

$55 \div \quad = 11$ $40 \div \quad = 5$ $84 \div \quad = 7$ $36 \div \quad = 6$ $60 \div \quad = 6$

$120 \div \quad = 12$ $48 \div \quad = 12$ $81 \div \quad = 9$ $24 \div \quad = 4$

NAME: _____   DATE: _____

BEAT THE DIVIDING WALL

23

© 2015 Tony Colledge

<u>KEY FOCUS</u>: *Dividing by 2 to 12*

How many can you get right ?

How fast can you do it ?

What will you try to get next time ?

YOU CAN DO IT !

SCORE	/37
mins	secs

TARGETS	/37
mins	secs

$72 \div = 9$

$132 \div = 12$ $56 \div = 8$

$36 \div = 6$ $63 \div = 7$ $24 \div = 2$

$84 \div = 12$ $27 \div = 3$ $42 \div = 7$ $120 \div = 10$

$48 \div = 4$ $25 \div = 5$ $96 \div = 8$ $28 \div = 4$ $54 \div = 6$

$60 \div = 12$ $24 \div = 6$ $121 \div = 11$ $40 \div = 8$

$44 \div = 11$ $81 \div = 9$ $36 \div = 12$ $45 \div = 5$ $72 \div = 6$

$108 \div = 12$ $35 \div = 5$ $64 \div = 8$ $21 \div = 7$

$55 \div = 11$ $36 \div = 9$ $144 \div = 12$ $24 \div = 3$ $49 \div = 7$

$48 \div = 6$ $77 \div = 7$ $32 \div = 8$ $60 \div = 10$

NAME: _____   DATE: _____

BEAT THE
DIVIDING WALL

KEY FOCUS: *Dividing by 2 to 12*

How many can you get right ?

How fast can you do it ?

What will you try to get next time ?

YOU CAN DO IT !

24		
SCORE		/37
mins		secs
TARGETS		/37
mins		secs

$108 \div \quad = 9$

$24 \div \quad = 12$ $56 \div \quad = 7$

$84 \div \quad = 7$ $36 \div \quad = 6$ $144 \div \quad = 12$

$48 \div \quad = 8$ $77 \div \quad = 11$ $32 \div \quad = 4$ $54 \div \quad = 9$

$25 \div \quad = 5$ $96 \div \quad = 12$ $45 \div \quad = 9$ $60 \div \quad = 5$ $44 \div \quad = 4$

$48 \div \quad = 12$ $64 \div \quad = 8$ $21 \div \quad = 3$ $121 \div \quad = 11$

$36 \div \quad = 4$ $81 \div \quad = 9$ $42 \div \quad = 6$ $72 \div \quad = 12$ $27 \div \quad = 9$

$132 \div \quad = 11$ $35 \div \quad = 7$ $60 \div \quad = 6$ $24 \div \quad = 8$

$40 \div \quad = 5$ $72 \div \quad = 8$ $24 \div \quad = 4$ $49 \div \quad = 7$ $36 \div \quad = 3$

$63 \div \quad = 9$ $55 \div \quad = 5$ $28 \div \quad = 7$ $120 \div \quad = 12$

NAME: _____ DATE: _____

BEAT THE DIVIDING WALL

© 2015 Tony Colledge   __KEY FOCUS__: *Dividing by 2 to 12*

25

How many can you get right ?

How fast can you do it ?

What will you try to get next time ?

YOU CAN DO IT !

SCORE	/37
mins	secs

TARGETS	/37
mins	secs

72 ÷   = 6

54 ÷   = 6   28 ÷   = 4

66 ÷   = 11   24 ÷   = 3   108 ÷   = 12

120 ÷   = 10   48 ÷   = 6   63 ÷   = 7   21 ÷   = 7

56 ÷   = 8   32 ÷   = 8   96 ÷   = 8   60 ÷   = 12   42 ÷   = 7

40 ÷   = 8   99 ÷   = 9   35 ÷   = 5   132 ÷   = 12

81 ÷   = 9   48 ÷   = 4   36 ÷   = 12   72 ÷   = 9   24 ÷   = 2

144 ÷   = 12   24 ÷   = 6   60 ÷   = 10   36 ÷   = 6

45 ÷   = 5   25 ÷   = 5   84 ÷   = 12   27 ÷   = 3   88 ÷   = 11

64 ÷   = 8   36 ÷   = 9   49 ÷   = 7   121 ÷   = 11

NAME: _____ DATE: _____

BEAT THE DIVIDING WALL

26

<u>KEY FOCUS</u>: *Dividing by 2 to 12 using factor pairs*

How many can you get right ?

How fast can you do it ?

What will you try to get next time ?

YOU CAN DO IT !

SCORE	/42
mins	secs
TARGETS	/42
mins	secs

45 ÷   =

99 ÷   =   96 ÷   =

132 ÷   =   81 ÷   =   32 ÷   =

60 ÷   =
60 ÷   =   63 ÷   =   21 ÷   =   72 ÷   =
72 ÷   =

144 ÷   =   56 ÷   =   42 ÷   =   84 ÷   =   90 ÷   =

36 ÷   =
36 ÷   =   36 ÷   =   24 ÷   =   24 ÷   =
24 ÷   =

88 ÷   =   28 ÷   =   48 ÷   =
48 ÷   =   35 ÷   =   121 ÷   =

55 ÷   =   108 ÷   =   64 ÷   =   9 ÷   =

14 ÷   =   54 ÷   =   44 ÷   =   77 ÷   =   25 ÷   =

27 ÷   =   49 ÷   =   15 ÷   =   120 ÷   =

NAME: _____   DATE: _____

BEAT THE DIVIDING WALL

27

KEY FOCUS: *Dividing by 2 to 12 using factor pairs*

How many can you get right ?

How fast can you do it ?

What will you try to get next time ?

YOU CAN DO IT !

SCORE	/42
mins	secs
TARGETS	/42
mins	secs

$56 \div =$

$84 \div =$ $108 \div =$

$25 \div =$ $72 \div =$ $81 \div =$
$72 \div =$

$45 \div =$ $96 \div =$ $15 \div =$ $121 \div =$

$88 \div =$ $35 \div =$ $132 \div =$ $54 \div =$ $14 \div =$

$24 \div =$ $24 \div =$ $99 \div =$ $48 \div =$
$24 \div =$ $48 \div =$

$9 \div =$ $64 \div =$ $55 \div =$ $144 \div =$ $32 \div =$

$27 \div =$ $77 \div =$ $42 \div =$ $120 \div =$

$63 \div =$ $44 \div =$ $90 \div =$ $49 \div =$ $21 \div =$

$36 \div =$ $36 \div =$ $28 \div =$ $60 \div =$
$36 \div =$ $60 \div =$

NAME: _____   DATE: _____

BEAT THE DIVIDING WALL

© 2015 Tony Colledge

28

KEY FOCUS: *Dividing by 2 to 12 using factor pairs*

How many can you get right ?

How fast can you do it ?

What will you try to get next time ?

YOU CAN DO IT !

SCORE	/42
mins	secs
TARGETS	/42
mins	secs

$121 \div \quad =$

$35 \div \quad =$ $96 \div \quad =$

$120 \div \quad =$ $54 \div \quad =$ $42 \div \quad =$

$81 \div \quad =$ $63 \div \quad =$ $99 \div \quad =$ $144 \div \quad =$

$33 \div \quad =$ $28 \div \quad =$ $84 \div \quad =$ $49 \div \quad =$ $27 \div \quad =$

$60 \div \quad =$
$60 \div \quad =$ $9 \div \quad =$ $24 \div \quad =$ $24 \div \quad =$
$24 \div \quad =$

$132 \div \quad =$ $15 \div \quad =$ $48 \div \quad =$
$48 \div \quad =$ $56 \div \quad =$ $25 \div \quad =$

$36 \div \quad =$
$36 \div \quad =$ $36 \div \quad =$ $88 \div \quad =$ $72 \div \quad =$
$72 \div \quad =$

$108 \div \quad =$ $45 \div \quad =$ $66 \div \quad =$ $21 \div \quad =$ $44 \div \quad =$

$14 \div \quad =$ $90 \div \quad =$ $64 \div \quad =$ $32 \div \quad =$

BEAT THE DIVIDING WALL

29

KEY FOCUS: *Dividing by 2 to 12 using factor pairs*

How many can you get right ?

How fast can you do it ?

What will you try to get next time ?

YOU CAN DO IT !

SCORE	/42
mins	secs
TARGETS	/42
mins	secs

$28 \div \quad =$

$99 \div \quad =$ $64 \div \quad =$

$54 \div \quad =$ $60 \div \quad =$ $108 \div \quad =$
                      $60 \div \quad =$

$24 \div \quad =$ $24 \div \quad =$ $36 \div \quad =$ $36 \div \quad =$
$24 \div \quad =$ $36 \div \quad =$

$90 \div \quad =$ $55 \div \quad =$ $132 \div \quad =$ $42 \div \quad =$ $96 \div \quad =$

$72 \div \quad =$ $49 \div \quad =$ $88 \div \quad =$ $48 \div \quad =$
$72 \div \quad =$ $48 \div \quad =$

$81 \div \quad =$ $120 \div \quad =$ $27 \div \quad =$ $35 \div \quad =$ $63 \div \quad =$

$25 \div \quad =$ $56 \div \quad =$ $121 \div \quad =$ $21 \div \quad =$

$15 \div \quad =$ $77 \div \quad =$ $9 \div \quad =$ $84 \div \quad =$ $14 \div \quad =$

$32 \div \quad =$ $44 \div \quad =$ $144 \div \quad =$ $45 \div \quad =$

NAME: _____   DATE: _____

BEAT THE DIVIDING WALL

KEY FOCUS: *Dividing by 2 to 12 using factor pairs*

30

How many can you get right ?

How fast can you do it ?

What will you try to get next time ?

YOU CAN DO IT !

SCORE	/42
mins	secs
TARGETS	/42
mins	secs

$63 \div \quad =$

$120 \div \quad =$ $81 \div \quad =$

$35 \div \quad =$ $48 \div \quad =$
$48 \div \quad =$ $88 \div \quad =$

$56 \div \quad =$ $99 \div \quad =$ $28 \div \quad =$ $132 \div \quad =$

$84 \div \quad =$ $60 \div \quad =$
$60 \div \quad =$ $64 \div \quad =$ $72 \div \quad =$
$72 \div \quad =$ $25 \div \quad =$

$42 \div \quad =$ $96 \div \quad =$ $55 \div \quad =$ $121 \div \quad =$

$24 \div \quad =$
$24 \div \quad =$ $24 \div \quad =$ $77 \div \quad =$ $36 \div \quad =$ $36 \div \quad =$
$36 \div \quad =$

$108 \div \quad =$ $54 \div \quad =$ $90 \div \quad =$ $144 \div \quad =$

$21 \div \quad =$ $9 \div \quad =$ $32 \div \quad =$ $45 \div \quad =$ $14 \div \quad =$

$15 \div \quad =$ $66 \div \quad =$ $49 \div \quad =$ $27 \div \quad =$

Beat The Dividing Wall

ANSWERS

DATE	TASK	SCORE	TIME	TARGET(S)

ANSWERS
BEAT THE DIVIDING WALL
© 2015 Tony Colledge

2 — Total 37

What was the final score ?

How fast was it completed ?

Were the previous targets beaten ?

What are the targets for next time ?

$100 \div 10 = 10$

$90 \div 10 = 9$ | $81 \div 9 = 9$

$80 \div 8 = 10$ | $72 \div 8 = 9$ | $70 \div 10 = 7$

$64 \div 8 = 8$ | $63 \div 9 = 7$ | $60 \div 6 = 10$ | $56 \div 7 = 8$

$54 \div 6 = 9$ | $50 \div 10 = 5$ | $49 \div 7 = 7$ | $48 \div 8 = 6$ | $45 \div 9 = 5$

$42 \div 6 = 7$ | $40 \div 8 = 5$ | $36 \div 9 = 4$ | $35 \div 5 = 7$

$32 \div 4 = 8$ | $30 \div 3 = 10$ | $28 \div 7 = 4$ | $27 \div 3 = 9$ | $25 \div 5 = 5$

$24 \div 3 = 8$ | $21 \div 3 = 7$ | $20 \div 5 = 4$ | $18 \div 6 = 3$

$16 \div 8 = 2$ | $15 \div 5 = 3$ | $14 \div 7 = 2$ | $12 \div 3 = 4$ | $10 \div 2 = 5$

$9 \div 3 = 3$ | $8 \div 2 = 4$ | $6 \div 3 = 2$ | $4 \div 2 = 2$

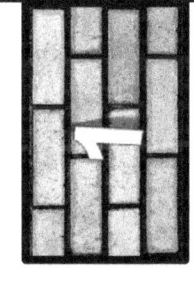

ANSWERS
BEAT THE DIVIDING WALL
© 2015 Tony Colledge

1 — Total 37

What was the final score ?

How fast was it completed ?

Were the previous targets beaten ?

What are the targets for next time ?

$4 \div 2 = 2$

$8 \div 4 = 2$ | $6 \div 2 = 3$

$12 \div 2 = 6$ | $10 \div 5 = 2$ | $9 \div 3 = 3$

$18 \div 9 = 2$ | $16 \div 4 = 4$ | $15 \div 3 = 5$ | $14 \div 2 = 7$

$27 \div 9 = 3$ | $25 \div 5 = 5$ | $24 \div 4 = 6$ | $21 \div 7 = 3$ | $20 \div 2 = 10$

$35 \div 7 = 5$ | $32 \div 8 = 4$ | $30 \div 5 = 6$ | $28 \div 4 = 7$

$48 \div 6 = 8$ | $45 \div 5 = 9$ | $42 \div 7 = 6$ | $40 \div 10 = 4$ | $36 \div 6 = 6$

$56 \div 8 = 7$ | $54 \div 9 = 6$ | $50 \div 5 = 10$ | $49 \div 7 = 7$

$72 \div 9 = 8$ | $70 \div 7 = 10$ | $64 \div 8 = 8$ | $63 \div 7 = 9$ | $60 \div 10 = 6$

$100 \div 10 = 10$ | $90 \div 9 = 10$ | $81 \div 9 = 9$ | $80 \div 10 = 8$

ANSWERS

BEAT THE DIVIDING WALL

© 2015 Tony Colledge

Total 37

4

What was the final score ?

How fast was it completed ?

Were the previous targets beaten ?

What are the targets for next time ?

$54 \div 6 = 9$

$28 \div 7 = 4$ $45 \div 9 = 5$

$35 \div 5 = 7$ $56 \div 7 = 8$ $12 \div 4 = 3$

$72 \div 8 = 9$ $18 \div 3 = 6$ $40 \div 5 = 8$ $27 \div 3 = 9$

$30 \div 10 = 3$ $42 \div 6 = 7$ $4 \div 2 = 2$ $100 \div 10 = 10$ $21 \div 3 = 7$

$60 \div 6 = 10$ $16 \div 2 = 8$ $49 \div 7 = 7$ $9 \div 3 = 3$

$64 \div 8 = 8$ $10 \div 2 = 5$ $48 \div 8 = 6$ $25 \div 5 = 5$ $80 \div 8 = 10$

$6 \div 3 = 2$ $50 \div 10 = 5$ $14 \div 7 = 2$ $32 \div 4 = 8$

$36 \div 4 = 9$ $15 \div 5 = 3$ $90 \div 10 = 9$ $20 \div 4 = 5$ $63 \div 9 = 7$

$24 \div 8 = 3$ $70 \div 10 = 7$ $8 \div 2 = 4$ $81 \div 9 = 9$

ANSWERS

BEAT THE DIVIDING WALL

© 2015 Tony Colledge

Total 37

3

What was the final score ?

How fast was it completed ?

Were the previous targets beaten ?

What are the targets for next time ?

$63 \div 7 = 9$

$64 \div 8 = 8$ $42 \div 7 = 6$

$16 \div 4 = 4$ $60 \div 10 = 6$ $9 \div 3 = 3$

$45 \div 5 = 9$ $24 \div 6 = 4$ $36 \div 6 = 6$ $72 \div 9 = 8$ $20 \div 10 = 2$

$27 \div 9 = 3$ $4 \div 2 = 2$ $100 \div 10 = 10$ $12 \div 6 = 2$ $25 \div 5 = 5$ $56 \div 8 = 7$

$10 \div 5 = 2$ $54 \div 9 = 6$ $30 \div 6 = 5$ $70 \div 7 = 10$ $49 \div 7 = 7$

$90 \div 9 = 10$ $32 \div 8 = 4$ $15 \div 3 = 5$ $48 \div 6 = 8$ $18 \div 2 = 9$

$81 \div 9 = 9$ $14 \div 2 = 7$ $80 \div 10 = 8$ $6 \div 2 = 3$

$50 \div 5 = 10$ $35 \div 7 = 5$ $40 \div 4 = 10$ $28 \div 4 = 7$

$8 \div 4 = 2$ $21 \div 7 = 3$

Top sheet

ANSWERS
BEAT THE DIVIDING WALL

Total 45

What was the final score ?

How fast was it completed ?

Were the previous targets beaten ?

What are the targets for next time ?

- $72 \div 8 = 9$
- $49 \div 7 = 7$
- $32 \div 4 = 8$
- $90 \div 10 = 9$
- $10 \div 2 = 5$
- $64 \div 8 = 8$
- $20 \div 5 = 4$
- $20 \div 2 = 10$
- $27 \div 3 = 9$
- $56 \div 7 = 8$
- $48 \div 8 = 6$
- $6 \div 3 = 2$
- $81 \div 9 = 9$
- $12 \div 2 = 6$
- $12 \div 3 = 4$
- $24 \div 3 = 8$
- $24 \div 4 = 6$
- $54 \div 6 = 9$
- $14 \div 7 = 2$
- $8 \div 2 = 4$
- $63 \div 9 = 7$
- $60 \div 6 = 10$
- $18 \div 9 = 2$
- $18 \div 6 = 3$
- $28 \div 7 = 4$
- $80 \div 8 = 10$
- $30 \div 3 = 10$
- $30 \div 5 = 6$
- $9 \div 3 = 3$
- $4 \div 2 = 2$
- $35 \div 5 = 7$
- $36 \div 9 = 4$
- $36 \div 6 = 6$
- $45 \div 9 = 5$
- $50 \div 10 = 5$
- $21 \div 3 = 7$
- $70 \div 10 = 7$
- $25 \div 5 = 5$
- $40 \div 10 = 4$
- $40 \div 8 = 5$
- $42 \div 6 = 7$
- $100 \div 10 = 10$
- $16 \div 4 = 4$
- $16 \div 8 = 2$

Bottom sheet

ANSWERS
BEAT THE DIVIDING WALL

Total 37

What was the final score ?

How fast was it completed ?

Were the previous targets beaten ?

What are the targets for next time ?

- $81 \div 9 = 9$
- $48 \div 6 = 8$
- $15 \div 3 = 5$
- $45 \div 5 = 9$
- $80 \div 10 = 8$
- $63 \div 7 = 9$
- $72 \div 9 = 8$
- $28 \div 4 = 7$
- $27 \div 9 = 3$
- $36 \div 9 = 4$
- $40 \div 10 = 4$
- $100 \div 10 = 10$
- $54 \div 9 = 6$
- $64 \div 8 = 8$
- $9 \div 3 = 3$
- $6 \div 2 = 3$
- $90 \div 9 = 10$
- $14 \div 2 = 7$
- $12 \div 2 = 6$
- $21 \div 7 = 3$
- $42 \div 7 = 6$
- $70 \div 7 = 10$
- $56 \div 8 = 7$
- $4 \div 2 = 2$
- $24 \div 4 = 6$
- $50 \div 5 = 10$
- $18 \div 9 = 2$
- $16 \div 4 = 4$
- $8 \div 4 = 2$
- $20 \div 2 = 10$
- $35 \div 7 = 5$
- $49 \div 7 = 7$
- $25 \div 5 = 5$
- $30 \div 5 = 6$
- $32 \div 8 = 4$
- $10 \div 5 = 2$
- $60 \div 10 = 6$

ANSWERS

BEAT THE DIVIDING WALL

8

Total 45

What was the final score ?

How fast was it completed ?

Were the previous targets beaten ?

What are the targets for next time ?

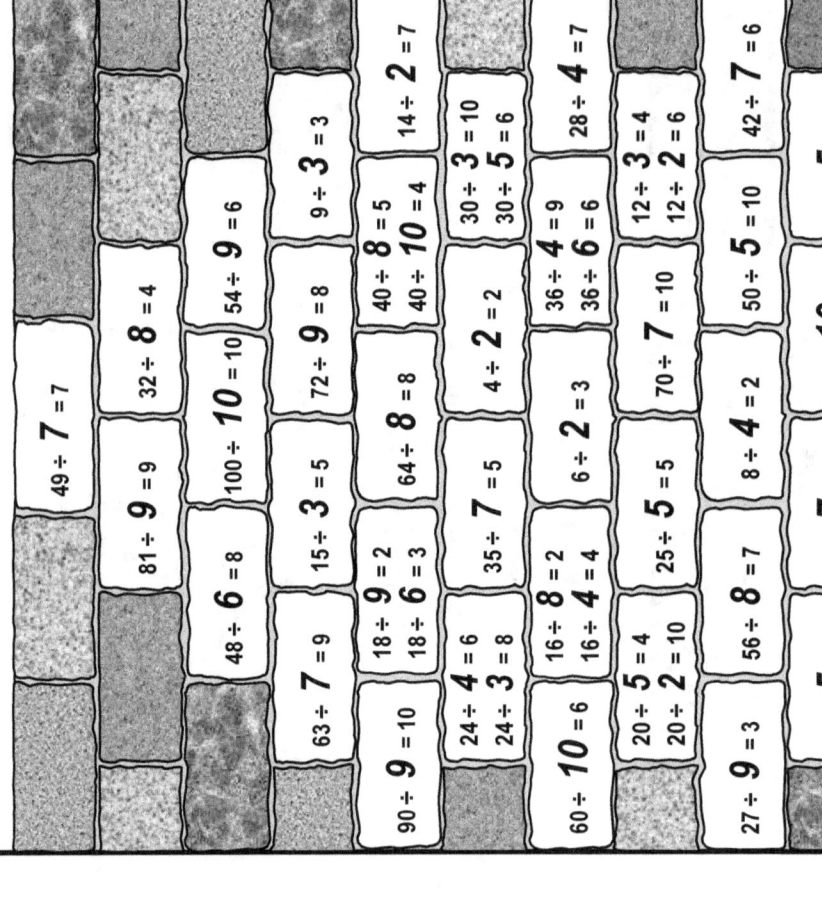

49 ÷ 7 = 7				
81 ÷ 9 = 9	32 ÷ 8 = 4			
48 ÷ 6 = 8	100 ÷ 10 = 10	54 ÷ 9 = 6		
63 ÷ 7 = 9	15 ÷ 3 = 5	72 ÷ 9 = 8	9 ÷ 3 = 3	
90 ÷ 9 = 10	18 ÷ 9 = 2 / 18 ÷ 6 = 3	64 ÷ 8 = 8	40 ÷ 8 = 5 / 40 ÷ 10 = 4	14 ÷ 2 = 7
24 ÷ 4 = 6 / 24 ÷ 3 = 8	35 ÷ 7 = 5	4 ÷ 2 = 2	30 ÷ 3 = 10 / 30 ÷ 5 = 6	
60 ÷ 10 = 6	16 ÷ 8 = 2 / 16 ÷ 4 = 4	6 ÷ 2 = 3	36 ÷ 4 = 9 / 36 ÷ 6 = 6	28 ÷ 4 = 7
20 ÷ 5 = 4 / 20 ÷ 2 = 10	25 ÷ 5 = 5	70 ÷ 7 = 10	12 ÷ 3 = 4 / 12 ÷ 2 = 6	
27 ÷ 9 = 3	56 ÷ 8 = 7	8 ÷ 4 = 2	50 ÷ 5 = 10	42 ÷ 7 = 6
45 ÷ 5 = 9	21 ÷ 7 = 3	80 ÷ 10 = 8	10 ÷ 5 = 2	

ANSWERS

BEAT THE DIVIDING WALL

7

Total 45

What was the final score ?

How fast was it completed ?

Were the previous targets beaten ?

What are the targets for next time ?

56 ÷ 7 = 8				
45 ÷ 9 = 5	90 ÷ 10 = 9			
40 ÷ 4 = 10 / 40 ÷ 5 = 8	63 ÷ 9 = 7	24 ÷ 8 = 3 / 24 ÷ 6 = 4		
48 ÷ 8 = 6	27 ÷ 3 = 9	15 ÷ 5 = 3	49 ÷ 7 = 7	
42 ÷ 6 = 7	36 ÷ 4 = 9 / 36 ÷ 6 = 6	9 ÷ 3 = 3	16 ÷ 4 = 4 / 16 ÷ 2 = 8	72 ÷ 8 = 9
80 ÷ 8 = 10	14 ÷ 7 = 2	50 ÷ 10 = 5	35 ÷ 5 = 7	
25 ÷ 5 = 5	12 ÷ 4 = 3 / 12 ÷ 6 = 2	64 ÷ 8 = 8	20 ÷ 10 = 2 / 20 ÷ 4 = 5	100 ÷ 10 = 10
6 ÷ 3 = 2	28 ÷ 7 = 4	70 ÷ 10 = 7	18 ÷ 3 = 6 / 18 ÷ 2 = 9	8 ÷ 2 = 4
81 ÷ 9 = 9	30 ÷ 10 = 3 / 30 ÷ 6 = 5	4 ÷ 2 = 2	60 ÷ 6 = 10	
21 ÷ 3 = 7	54 ÷ 6 = 9	32 ÷ 4 = 8	10 ÷ 2 = 5	

ANSWERS
BEAT THE DIVIDING WALL

10 Total 45

What was the final score ?

How fast was it completed ?

Were the previous targets beaten ?

What are the targets for next time ?

42 ÷ 7 = 6

64 ÷ 8 = 8   27 ÷ 9 = 3

15 ÷ 3 = 5   80 ÷ 10 = 8   56 ÷ 8 = 7

20 ÷ 5 = 4 / 20 ÷ 2 = 10   4 ÷ 2 = 2   54 ÷ 9 = 6   28 ÷ 4 = 7   18 ÷ 9 = 2 / 18 ÷ 6 = 3

50 ÷ 5 = 10   14 ÷ 2 = 7   72 ÷ 9 = 8

63 ÷ 7 = 9   12 ÷ 3 = 4 / 12 ÷ 2 = 6   30 ÷ 5 = 6 / 30 ÷ 3 = 10   100 ÷ 10 = 10

10 ÷ 5 = 2   32 ÷ 8 = 4   6 ÷ 2 = 3   70 ÷ 7 = 10

35 ÷ 7 = 5   36 ÷ 6 = 6 / 36 ÷ 9 = 4   16 ÷ 8 = 2 / 16 ÷ 4 = 4   81 ÷ 9 = 9

21 ÷ 7 = 3   48 ÷ 6 = 8   25 ÷ 5 = 5   9 ÷ 3 = 3

90 ÷ 9 = 10   24 ÷ 3 = 8 / 24 ÷ 4 = 6   8 ÷ 4 = 2   60 ÷ 10 = 6   40 ÷ 8 = 5 / 40 ÷ 10 = 4

ANSWERS
BEAT THE DIVIDING WALL

9 Total 45

What was the final score ?

How fast was it completed ?

Were the previous targets beaten ?

What are the targets for next time ?

48 ÷ 8 = 6

54 ÷ 6 = 9   21 ÷ 3 = 7

32 ÷ 4 = 8   63 ÷ 9 = 7   28 ÷ 7 = 4

20 ÷ 4 = 5 / 20 ÷ 10 = 2   80 ÷ 8 = 10   70 ÷ 10 = 7   81 ÷ 9 = 9   36 ÷ 6 = 6 / 36 ÷ 4 = 9   56 ÷ 7 = 8   24 ÷ 8 = 3 / 24 ÷ 6 = 4   35 ÷ 5 = 7

8 ÷ 2 = 4   16 ÷ 2 = 8 / 16 ÷ 4 = 4   45 ÷ 9 = 5   9 ÷ 3 = 3   40 ÷ 4 = 10 / 40 ÷ 5 = 8   6 ÷ 3 = 2

42 ÷ 6 = 7   25 ÷ 5 = 5   72 ÷ 8 = 9   60 ÷ 6 = 10   64 ÷ 8 = 8   27 ÷ 3 = 9   18 ÷ 3 = 6 / 18 ÷ 2 = 9

100 ÷ 10 = 10   15 ÷ 5 = 3   30 ÷ 10 = 3 / 30 ÷ 6 = 5   50 ÷ 10 = 5   49 ÷ 7 = 7

12 ÷ 4 = 3 / 12 ÷ 6 = 2   4 ÷ 2 = 2   90 ÷ 10 = 9   10 ÷ 2 = 5   14 ÷ 7 = 2

ANSWERS

BEAT THE DIVIDING WALL

12

Total 37

- What was the final score ?
- How fast was it completed ?
- Were the previous targets beaten ?
- What are the targets for next time ?

$63 \div 9 = 7$
$16 \div 8 = 2$
$36 \div 4 = 9$
$72 \div 8 = 9$
$42 \div 7 = 6$
$18 \div 9 = 2$
$54 \div 9 = 6$
$48 \div 8 = 6$
$25 \div 5 = 5$
$40 \div 4 = 10$
$81 \div 9 = 9$
$14 \div 2 = 7$
$27 \div 3 = 9$
$64 \div 8 = 8$
$70 \div 10 = 7$
$50 \div 10 = 5$
$56 \div 8 = 7$
$15 \div 5 = 3$
$8 \div 4 = 2$
$32 \div 4 = 8$
$9 \div 3 = 3$
$20 \div 5 = 4$
$60 \div 6 = 10$
$100 \div 10 = 10$
$6 \div 2 = 3$
$28 \div 7 = 4$
$45 \div 5 = 9$
$10 \div 2 = 5$
$30 \div 3 = 10$
$24 \div 4 = 6$
$80 \div 10 = 8$
$35 \div 5 = 7$
$21 \div 7 = 3$
$90 \div 10 = 9$
$49 \div 7 = 7$
$4 \div 2 = 2$
$12 \div 3 = 4$

ANSWERS

BEAT THE DIVIDING WALL

11

Total 37

- What was the final score ?
- How fast was it completed ?
- Were the previous targets beaten ?
- What are the targets for next time ?

$64 \div 8 = 8$
$27 \div 9 = 3$
$49 \div 7 = 7$
$63 \div 7 = 9$
$25 \div 5 = 5$
$48 \div 6 = 8$
$28 \div 4 = 7$
$6 \div 2 = 3$
$24 \div 3 = 8$
$45 \div 5 = 9$
$10 \div 2 = 5$
$32 \div 8 = 4$
$90 \div 9 = 10$
$21 \div 7 = 3$
$40 \div 8 = 5$
$100 \div 10 = 10$
$12 \div 6 = 2$
$56 \div 8 = 7$
$30 \div 5 = 6$
$18 \div 6 = 3$
$4 \div 2 = 2$
$20 \div 2 = 10$
$36 \div 6 = 6$
$42 \div 7 = 6$
$35 \div 7 = 5$
$8 \div 4 = 2$
$60 \div 10 = 6$
$9 \div 3 = 3$
$14 \div 2 = 7$
$72 \div 9 = 8$
$54 \div 9 = 6$
$15 \div 5 = 3$
$80 \div 10 = 8$
$81 \div 9 = 9$
$16 \div 4 = 4$
$50 \div 5 = 10$

ANSWERS

BEAT THE DIVIDING WALL

14

Total 37

What was the final score ?

How fast was it completed ?

Were the previous targets beaten ?

What are the targets for next time ?

54 ÷ 6 = 9

80 ÷ 8 = 10   28 ÷ 4 = 7

81 ÷ 9 = 9   30 ÷ 5 = 6   63 ÷ 7 = 9

42 ÷ 7 = 6   24 ÷ 4 = 6   12 ÷ 4 = 3   40 ÷ 8 = 5   16 ÷ 4 = 4

6 ÷ 2 = 3   60 ÷ 10 = 6   45 ÷ 9 = 5   27 ÷ 9 = 3   16 ÷ 8 = 2

18 ÷ 9 = 2   25 ÷ 5 = 5   48 ÷ 6 = 8   21 ÷ 7 = 3

72 ÷ 9 = 8   90 ÷ 10 = 9   4 ÷ 2 = 2

50 ÷ 5 = 10   15 ÷ 3 = 5   56 ÷ 7 = 8   70 ÷ 7 = 10

100 ÷ 10 = 10   35 ÷ 7 = 5   32 ÷ 8 = 4   49 ÷ 7 = 7

20 ÷ 4 = 5   9 ÷ 3 = 3   14 ÷ 7 = 2   36 ÷ 9 = 4   64 ÷ 8 = 8

ANSWERS

BEAT THE DIVIDING WALL

13

Total 37

What was the final score ?

How fast was it completed ?

Were the previous targets beaten ?

What are the targets for next time ?

81 ÷ 9 = 9

40 ÷ 8 = 5   56 ÷ 7 = 8

49 ÷ 7 = 7   27 ÷ 9 = 3   42 ÷ 6 = 7

60 ÷ 10 = 6   54 ÷ 6 = 9   48 ÷ 6 = 8   20 ÷ 4 = 5   28 ÷ 7 = 4

25 ÷ 5 = 5   70 ÷ 7 = 10   6 ÷ 3 = 2   64 ÷ 8 = 8   8 ÷ 2 = 4   10 ÷ 2 = 5

72 ÷ 9 = 8   45 ÷ 9 = 5   50 ÷ 5 = 10   32 ÷ 8 = 4   15 ÷ 5 = 3

9 ÷ 3 = 3   21 ÷ 7 = 3   90 ÷ 10 = 9   14 ÷ 2 = 7   4 ÷ 2 = 2

100 ÷ 10 = 10   24 ÷ 3 = 8   63 ÷ 9 = 7   30 ÷ 10 = 3

35 ÷ 5 = 7   16 ÷ 8 = 2   80 ÷ 8 = 10   18 ÷ 2 = 9

36 ÷ 6 = 6   12 ÷ 4 = 3

ANSWERS

BEAT THE DIVIDING WALL — 15

© 2015 Tony Colledge

What was the final score ?

How fast was it completed ?

Were the previous targets beaten ?

What are the targets for next time ?

Total 37

$64 \div 8 = 8$

$27 \div 9 = 3$ | $45 \div 5 = 9$

$72 \div 8 = 9$ | $56 \div 8 = 7$ | $100 \div 10 = 10$

$9 \div 3 = 3$ | $42 \div 6 = 7$ | $32 \div 4 = 8$ | $36 \div 9 = 4$ | $25 \div 5 = 5$

$80 \div 10 = 8$ | $20 \div 5 = 4$ | $24 \div 8 = 3$ | $60 \div 10 = 6$ | $63 \div 7 = 9$

$50 \div 5 = 10$ | $40 \div 5 = 8$ | $12 \div 6 = 2$ | $49 \div 7 = 7$ | $30 \div 5 = 6$

$14 \div 2 = 7$ | $81 \div 9 = 9$ | $90 \div 9 = 10$ | $15 \div 3 = 5$

$35 \div 5 = 7$ | $16 \div 4 = 4$ | $6 \div 2 = 3$ | $54 \div 6 = 9$ | $28 \div 7 = 4$

$48 \div 8 = 6$ | $4 \div 2 = 2$ | $10 \div 5 = 2$ | $18 \div 3 = 6$ | $8 \div 4 = 2$

$21 \div 3 = 7$ | $70 \div 7 = 10$

ANSWERS

BEAT THE DIVIDING WALL — 16

© 2015 Tony Colledge

What was the final score ?

How fast was it completed ?

Were the previous targets beaten ?

What are the targets for next time ?

Total 45

$54 \div 6,9$

$72 \div 8,9$ | $48 \div 6,8$

$100 \div 10,10$ | $64 \div 8,8$ | $35 \div 5,7$

$36 \div 4,9$ / $36 \div 6,6$ | $15 \div 3,5$ | $56 \div 7,8$ | $30 \div 3,10$ / $30 \div 5,6$

$28 \div 4,7$ | $25 \div 5,5$ | $90 \div 9,10$ | $4 \div 2,2$ | $49 \div 7,7$

$12 \div 2,6$ / $12 \div 3,4$ | $50 \div 5,10$ | $27 \div 3,9$ | $20 \div 2,10$ / $20 \div 4,5$

$9 \div 3,3$ | $70 \div 7,10$ | $32 \div 4,8$ | $45 \div 5,9$ | $21 \div 3,7$

$16 \div 2,8$ / $16 \div 4,4$ | $81 \div 9,9$ | $60 \div 6,10$ | $24 \div 3,8$ / $24 \div 4,6$

$8 \div 2,4$ | $42 \div 6,7$ | $14 \div 2,7$ | $80 \div 8,10$ | $6 \div 2,3$

$40 \div 4,10$ / $40 \div 5,8$ | $63 \div 7,9$ | $10 \div 2,5$ | $18 \div 2,9$ / $18 \div 3,6$

ANSWERS
BEAT THE DIVIDING WALL

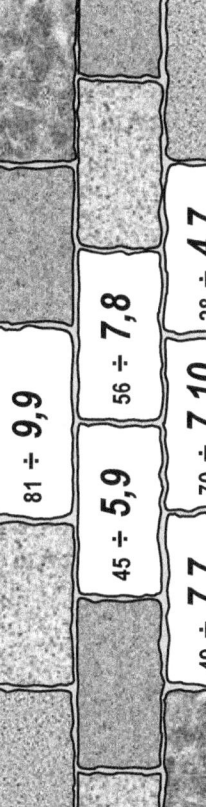

18

Total 45

- What was the final score ?
- How fast was it completed ?
- Were the previous targets beaten ?
- What are the targets for next time ?

81 ÷ 9,9

45 ÷ 5,9   56 ÷ 7,8

49 ÷ 7,7   70 ÷ 7,10   28 ÷ 4,7

10 ÷ 2,5   54 ÷ 6,9   8 ÷ 2,4   100 ÷ 10,10   4 ÷ 2,2

14 ÷ 2,7   20÷2,10 20÷4,5   32 ÷ 4,8   18÷2,9 18÷3,6   30÷3,10 30÷5,6   42 ÷ 6,7

9 ÷ 3,3   12÷2,6 12÷3,4   27 ÷ 3,9   80 ÷ 8,10

90 ÷ 9,10   15 ÷ 3,5   64 ÷ 8,8   36÷4,9 36÷6,6   50 ÷ 5,10

35 ÷ 5,7   16÷2,8 16÷4,4   72 ÷ 8,9   40÷4,10 40÷5,8   6 ÷ 2,3

21 ÷ 3,7   24÷3,8 24÷4,6   48 ÷ 6,8   25 ÷ 5,5   63 ÷ 7,9

60 ÷ 6,10

ANSWERS
BEAT THE DIVIDING WALL

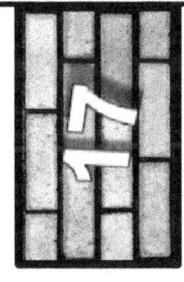

17

Total 45

- What was the final score ?
- How fast was it completed ?
- Were the previous targets beaten ?
- What are the targets for next time ?

28 ÷ 4,7

64 ÷ 8,8   54 ÷ 6,9

48 ÷ 6,8   9 ÷ 3,3   35 ÷ 5,7

32 ÷ 4,8   18÷2,9 18÷3,6   8 ÷ 2,4

14 ÷ 2,7   60 ÷ 6,10   42 ÷ 6,7   40÷4,10 40÷5,8   27 ÷ 3,9

16÷2,8 16÷4,4   12÷2,6 12÷3,4   10 ÷ 2,5   80 ÷ 8,10   36÷4,9 36÷6,6   56 ÷ 7,8

90 ÷ 9,10   49 ÷ 7,7   6 ÷ 2,3   70 ÷ 7,10   81 ÷ 9,9   50 ÷ 5,10

25 ÷ 5,5   20÷2,10 20÷4,5   21 ÷ 3,7   72 ÷ 8,9   45 ÷ 5,9

4 ÷ 2,2   30÷3,10 30÷5,6   15 ÷ 3,5   100 ÷ 10,10   24÷3,8 24÷4,6

ANSWERS

BEAT THE DIVIDING WALL

20

Total 45

What was the final score ?

How fast was it completed ?

Were the previous targets beaten ?

What are the targets for next time ?

$56 \div 7,8$

$64 \div 8,8$ | $100 \div 10,10$ | $28 \div 4,7$ | $81 \div 9,9$ | $15 \div 3,5$

$45 \div 5,9$ | $60 \div 6,10$ | $25 \div 5,5$ | $49 \div 7,7$ | $6 \div 2,3$

$14 \div 2,7$ | $32 \div 4,8$ | $70 \div 7,10$ | $63 \div 7,9$

$16 \div 2,8$ / $16 \div 4,4$ | $36 \div 4,9$ / $36 \div 6,6$ | $40 \div 4,10$ / $40 \div 5,8$ | $24 \div 3,8$ / $24 \div 4,6$

$42 \div 6,7$ | $80 \div 8,10$ | $72 \div 8,9$ | $8 \div 2,4$ | $50 \div 5,10$

$20 \div 2,10$ / $20 \div 4,5$ | $12 \div 2,6$ / $12 \div 3,4$ | $30 \div 3,10$ / $30 \div 5,6$ | $18 \div 2,9$ / $18 \div 3,6$

$27 \div 3,9$ | $10 \div 2,5$ | $35 \div 5,7$ | $9 \div 3,3$ | $54 \div 6,9$

$21 \div 3,7$ | $4 \div 2,2$ | $90 \div 9,10$ | $48 \div 6,8$

ANSWERS

BEAT THE DIVIDING WALL

19

Total 45

What was the final score ?

How fast was it completed ?

Were the previous targets beaten ?

What are the targets for next time ?

$27 \div 3,9$

$21 \div 3,7$ | $48 \div 6,8$ | $80 \div 8,10$ | $63 \div 7,9$ | $9 \div 3,3$ | $60 \div 6,10$

$42 \div 6,7$ | $54 \div 6,9$ | $45 \div 5,9$ | $28 \div 4,7$ | $12 \div 2,6$ / $12 \div 3,4$

$72 \div 8,9$ | $40 \div 4,10$ / $40 \div 5,8$ | $36 \div 4,9$ / $36 \div 6,6$ | $56 \div 7,8$ | $15 \div 3,5$ | $16 \div 2,8$ / $16 \div 4,4$ | $4 \div 2,2$

$25 \div 5,5$ | $90 \div 9,10$ | $32 \div 4,8$ | $81 \div 9,9$ | $50 \div 5,10$ | $18 \div 2,9$ / $18 \div 3,6$

$30 \div 3,10$ / $30 \div 5,6$ | $6 \div 2,3$ | $20 \div 2,10$ / $20 \div 4,5$ | $14 \div 2,7$

$70 \div 7,10$ | $24 \div 3,8$ / $24 \div 4,6$ | $10 \div 2,5$ | $64 \div 8,8$ | $35 \div 5,7$

$49 \div 7,7$ | $8 \div 2,4$

ANSWERS

BEAT THE DIVIDING WALL

22

Total 37

What was the final score ?

How fast was it completed ?

Were the previous targets beaten ?

What are the targets for next time ?

121 ÷ 11 = 11

63 ÷ 7 = 9   32 ÷ 8 = 4

45 ÷ 5 = 9   96 ÷ 8 = 12   24 ÷ 3 = 8

144 ÷ 12 = 12   21 ÷ 7 = 3   54 ÷ 6 = 9   36 ÷ 9 = 4

56 ÷ 8 = 7   42 ÷ 7 = 6   77 ÷ 7 = 11   48 ÷ 6 = 8

44 ÷ 11 = 4   72 ÷ 9 = 8   27 ÷ 3 = 9   132 ÷ 11 = 12

49 ÷ 7 = 7   28 ÷ 4 = 7   60 ÷ 12 = 5   25 ÷ 5 = 5

24 ÷ 2 = 12   64 ÷ 8 = 8   35 ÷ 5 = 7   72 ÷ 6 = 12

40 ÷ 8 = 5   84 ÷ 12 = 7   36 ÷ 6 = 6   60 ÷ 10 = 6

55 ÷ 5 = 11   120 ÷ 10 = 12   48 ÷ 4 = 12   81 ÷ 9 = 9   24 ÷ 6 = 4

ANSWERS

BEAT THE DIVIDING WALL

21

Total 37

What was the final score ?

How fast was it completed ?

Were the previous targets beaten ?

What are the targets for next time ?

144 ÷ 12 = 12

121 ÷ 11 = 11   132 ÷ 11 = 12

96 ÷ 12 = 8   108 ÷ 9 = 12   120 ÷ 12 = 10

72 ÷ 12 = 6   77 ÷ 11 = 7   81 ÷ 9 = 9   84 ÷ 7 = 12

60 ÷ 6 = 10   54 ÷ 9 = 6   63 ÷ 9 = 7   64 ÷ 8 = 8   72 ÷ 8 = 9

49 ÷ 7 = 7   55 ÷ 5 = 11   56 ÷ 7 = 8

60 ÷ 5 = 12   44 ÷ 4 = 11   45 ÷ 9 = 5   48 ÷ 8 = 6   48 ÷ 12 = 4

42 ÷ 6 = 7   36 ÷ 4 = 9   40 ÷ 5 = 8

36 ÷ 6 = 6   28 ÷ 7 = 4   36 ÷ 3 = 12   32 ÷ 4 = 8   35 ÷ 7 = 5

25 ÷ 5 = 5   27 ÷ 9 = 3   24 ÷ 8 = 3   24 ÷ 12 = 2

21 ÷ 3 = 7   24 ÷ 4 = 6

Sheet 24

ANSWERS

BEAT THE DIVIDING WALL

24

Total 37

What was the final score ?

How fast was it completed ?

Were the previous targets beaten ?

What are the targets for next time ?

$108 \div 12 = 9$
$24 \div 2 = 12$
$56 \div 8 = 7$
$84 \div 12 = 7$
$36 \div 6 = 6$
$144 \div 12 = 12$
$48 \div 6 = 8$
$77 \div 7 = 11$
$32 \div 8 = 4$
$54 \div 6 = 9$
$25 \div 5 = 5$
$96 \div 8 = 12$
$45 \div 5 = 9$
$60 \div 12 = 5$
$44 \div 11 = 4$
$48 \div 4 = 12$
$64 \div 8 = 8$
$21 \div 7 = 3$
$121 \div 11 = 11$
$36 \div 9 = 4$
$81 \div 9 = 9$
$42 \div 7 = 6$
$72 \div 6 = 12$
$27 \div 3 = 9$
$132 \div 12 = 11$
$35 \div 5 = 7$
$60 \div 10 = 6$
$24 \div 3 = 8$
$40 \div 8 = 5$
$72 \div 9 = 8$
$24 \div 6 = 4$
$49 \div 7 = 7$
$36 \div 12 = 3$
$63 \div 7 = 9$
$55 \div 11 = 5$
$28 \div 4 = 7$
$120 \div 10 = 12$

Sheet 23

ANSWERS

BEAT THE DIVIDING WALL

23

Total 37

What was the final score ?

How fast was it completed ?

Were the previous targets beaten ?

What are the targets for next time ?

$72 \div 8 = 9$
$132 \div 11 = 12$
$56 \div 7 = 8$
$36 \div 6 = 6$
$63 \div 9 = 7$
$24 \div 12 = 2$
$84 \div 7 = 12$
$27 \div 9 = 3$
$42 \div 6 = 7$
$120 \div 12 = 10$
$60 \div 5 = 12$
$96 \div 12 = 8$
$28 \div 7 = 4$
$54 \div 9 = 6$
$44 \div 4 = 11$
$24 \div 4 = 6$
$121 \div 11 = 11$
$40 \div 5 = 8$
$81 \div 9 = 9$
$36 \div 3 = 12$
$45 \div 9 = 5$
$72 \div 12 = 6$
$108 \div 9 = 12$
$35 \div 7 = 5$
$64 \div 8 = 8$
$21 \div 3 = 7$
$55 \div 5 = 11$
$36 \div 4 = 9$
$144 \div 12 = 12$
$24 \div 8 = 3$
$49 \div 7 = 7$
$48 \div 8 = 6$
$77 \div 11 = 7$
$32 \div 4 = 8$
$60 \div 6 = 10$

ANSWERS

BEAT THE DIVIDING WALL

26

Total
42

What was the final score ?

How fast was it completed ?

Were the previous targets beaten ?

What are the targets for next time ?

45 ÷ **5,9**

96 ÷ **8,12**

99 ÷ **9,11**

132 ÷ **11,12**   81 ÷ **9,9**   32 ÷ **4,8**

21 ÷ **3,7**   72 ÷ **6,12** / 72 ÷ **8,9**

60 ÷ **5,12** / 60 ÷ **6,10**   63 ÷ **7,9**   42 ÷ **6,7**   84 ÷ **7,12**   90 ÷ **9,10**

144 ÷ **12,12**   56 ÷ **7,8**   36 ÷ **3,12**   24 ÷ **2,12**   24 ÷ **3,8** / 24 ÷ **4,6**

36 ÷ **4,9** / 36 ÷ **6,6**   48 ÷ **4,12** / 48 ÷ **6,8**   108 ÷ **9,12**   35 ÷ **5,7**   121 ÷ **11,11**

88 ÷ **8,11**   28 ÷ **4,7**   64 ÷ **8,8**   9 ÷ **3,3**

55 ÷ **5,11**   44 ÷ **4,11**   77 ÷ **7,11**   25 ÷ **5,5**

14 ÷ **2,7**   54 ÷ **6,9**   49 ÷ **7,7**   15 ÷ **3,5**   120 ÷ **10,12**

27 ÷ **3,9**

ANSWERS

BEAT THE DIVIDING WALL

25

Total
37

What was the final score ?

How fast was it completed ?

Were the previous targets beaten ?

What are the targets for next time ?

72 ÷ **12** = 6

54 ÷ **9** = 6   28 ÷ **7** = 4

66 ÷ **6** = 11   24 ÷ **8** = 3   108 ÷ **9** = 12   21 ÷ **3** = 7   42 ÷ **6** = 7

120 ÷ **12** = 10   48 ÷ **8** = 6   63 ÷ **9** = 7   60 ÷ **5** = 12   132 ÷ **11** = 12   24 ÷ **12** = 2

40 ÷ **5** = 8   96 ÷ **12** = 8   35 ÷ **7** = 5   72 ÷ **8** = 9   88 ÷ **8** = 11

32 ÷ **4** = 8   99 ÷ **11** = 9   36 ÷ **3** = 12   24 ÷ **12** = 2

81 ÷ **9** = 9   48 ÷ **12** = 4   36 ÷ **6** = 6   27 ÷ **9** = 3   121 ÷ **11** = 11

56 ÷ **7** = 8   144 ÷ **12** = 12   24 ÷ **4** = 6   60 ÷ **6** = 10   84 ÷ **7** = 12

45 ÷ **9** = 5   25 ÷ **5** = 5   36 ÷ **4** = 9   49 ÷ **7** = 7

64 ÷ **8** = 8

ANSWERS

BEAT THE DIVIDING WALL

28

Total 42

What was the final score ?

How fast was it completed ?

Were the previous targets beaten ?

What are the targets for next time ?

$121 \div 11,11$

$35 \div 5,7$ $96 \div 8,12$

$120 \div 10,12$ $54 \div 6,9$ $42 \div 6,7$

$81 \div 9,9$ $63 \div 7,9$ $99 \div 9,11$ $144 \div 12,12$

$33 \div 3,11$ $28 \div 4,7$ $84 \div 7,12$ $49 \div 7,7$ $27 \div 3,9$

$60 \div 5,12$ / $60 \div 6,10$ $9 \div 3,3$ $24 \div 2,12$ $24 \div 3,8$ / $24 \div 4,6$

$132 \div 11,12$ $15 \div 3,5$ $48 \div 4,12$ / $48 \div 6,8$ $56 \div 7,8$ $25 \div 5,5$

$36 \div 4,9$ / $36 \div 6,6$ $36 \div 3,12$ $88 \div 8,11$ $72 \div 6,12$ / $72 \div 8,9$

$108 \div 9,12$ $45 \div 5,9$ $66 \div 6,11$ $21 \div 3,7$ $44 \div 4,11$

$14 \div 2,7$ $90 \div 9,10$ $64 \div 8,8$ $32 \div 4,8$

ANSWERS

BEAT THE DIVIDING WALL

27

Total 42

What was the final score ?

How fast was it completed ?

Were the previous targets beaten ?

What are the targets for next time ?

$56 \div 7,8$

$84 \div 7,12$ $108 \div 9,12$

$25 \div 5,5$ $72 \div 6,12$ / $72 \div 8,9$ $81 \div 9,9$

$45 \div 5,9$ $35 \div 5,7$ $132 \div 11,12$ $15 \div 3,5$ $121 \div 11,11$

$24 \div 3,8$ / $24 \div 4,6$ $96 \div 8,12$ $24 \div 2,12$ $99 \div 9,11$ $54 \div 6,9$ $14 \div 2,7$

$9 \div 3,3$ $64 \div 8,8$ $55 \div 5,11$ $144 \div 12,12$ $48 \div 4,12$ / $48 \div 6,8$ $32 \div 4,8$

$27 \div 3,9$ $77 \div 7,11$ $42 \div 6,7$ $120 \div 10,12$

$88 \div 8,11$ $44 \div 4,11$ $90 \div 9,10$ $49 \div 7,7$ $21 \div 3,7$

$63 \div 7,9$ $36 \div 3,12$ $28 \div 4,7$ $60 \div 5,12$ / $60 \div 6,10$

$36 \div 4,9$ / $36 \div 6,6$

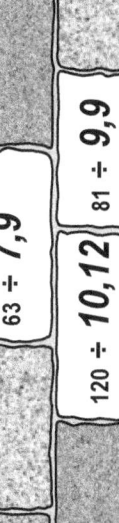

ANSWERS

BEAT THE DIVIDING WALL

30

Total 42

© 2015 Tony Colledge

What was the final score ?

How fast was it completed ?

Were the previous targets beaten ?

What are the targets for next time ?

63 ÷ 7,9
120 ÷ 10,12
81 ÷ 9,9
48 ÷ 4,12
48 ÷ 6,8
88 ÷ 8,11
35 ÷ 5,7
28 ÷ 4,7
132 ÷ 11,12
99 ÷ 9,11
64 ÷ 8,8
25 ÷ 5,5
56 ÷ 7,8
60 ÷ 5,12
60 ÷ 6,10
72 ÷ 6,12
72 ÷ 8,9
55 ÷ 5,11
121 ÷ 11,11
84 ÷ 7,12
42 ÷ 6,7
96 ÷ 8,12
77 ÷ 7,11
36 ÷ 3,12
36 ÷ 4,9
36 ÷ 6,6
24 ÷ 3,8
24 ÷ 4,6
24 ÷ 2,12
54 ÷ 6,9
90 ÷ 9,10
144 ÷ 12,12
108 ÷ 9,12
21 ÷ 3,7
9 ÷ 3,3
32 ÷ 4,8
45 ÷ 5,9
14 ÷ 2,7
15 ÷ 3,5
66 ÷ 6,11
49 ÷ 7,7
27 ÷ 3,9

ANSWERS

BEAT THE DIVIDING WALL

29

Total 42

© 2015 Tony Colledge

What was the final score ?

How fast was it completed ?

Were the previous targets beaten ?

What are the targets for next time ?

28 ÷ 4,7
64 ÷ 8,8
108 ÷ 9,12
36 ÷ 4,9
36 ÷ 6,6
96 ÷ 8,12
99 ÷ 9,11
60 ÷ 5,12
60 ÷ 6,10
36 ÷ 3,12
42 ÷ 6,7
48 ÷ 4,12
48 ÷ 6,8
63 ÷ 7,9
54 ÷ 6,9
24 ÷ 2,12
132 ÷ 11,12
88 ÷ 8,11
35 ÷ 5,7
21 ÷ 3,7
14 ÷ 2,7
49 ÷ 7,7
27 ÷ 3,9
121 ÷ 11,11
84 ÷ 7,12
45 ÷ 5,9
24 ÷ 3,8
24 ÷ 4,6
55 ÷ 5,11
72 ÷ 6,12
72 ÷ 8,9
120 ÷ 10,12
25 ÷ 5,5
56 ÷ 7,8
77 ÷ 7,11
9 ÷ 3,3
44 ÷ 4,11
90 ÷ 9,10
81 ÷ 9,9
15 ÷ 3,5
32 ÷ 4,8
144 ÷ 12,12

The Dividing Wall Extension

This section introduces division using the "bus shelter" layout

Sheets 1a to 1c cover mental division by 2 to 10
Sheets 2a to 2e cover mental division by 2 to 10, with remainders
Sheets 3a to 3c cover dividing numbers up to 999 by 2 to 9
Sheets 4a to 4c cover dividing numbers up to 9999 by 2 to 9
Sheets 5a and 5b cover dividing very large numbers by 2 to 9

Answers are provided at the end of this section.

NAME: _____   DATE: _____

THE DIVIDING WALL

KEY FOCUS: *Mentally dividing by 2 to 10*

Extension 1a

How did you do this time ?

SCORE	/ 40
mins	secs

TARGET(S)

What can you do to improve ?

Take the challenge – You can do it !

2 ⟌ 1 2	5 ⟌ 1 0	4 ⟌ 1 6	3 ⟌ 1 2	5 ⟌ 1 5
7 ⟌ 1 4	2 ⟌ 1 6	5 ⟌ 2 5	9 ⟌ 1 8	4 ⟌ 2 4
3 ⟌ 9	6 ⟌ 1 8	8 ⟌ 3 2	4 ⟌ 2 0	3 ⟌ 2 7
3 ⟌ 2 1	8 ⟌ 2 4	5 ⟌ 3 0	4 ⟌ 3 6	5 ⟌ 3 5
7 ⟌ 2 8	3 ⟌ 3 0	9 ⟌ 5 4	8 ⟌ 4 0	6 ⟌ 4 2
6 ⟌ 3 6	9 ⟌ 4 5	10 ⟌ 5 0	7 ⟌ 4 9	9 ⟌ 7 2
4 ⟌ 4 0	7 ⟌ 7 0	8 ⟌ 4 8	10 ⟌ 8 0	7 ⟌ 6 3
9 ⟌ 8 1	7 ⟌ 5 6	6 ⟌ 6 0	8 ⟌ 6 4	10 ⟌ 9 0

NAME: _____

DATE: _____

THE DIVIDING WALL

Extension **1b**

KEY FOCUS: *Mentally dividing by 2 to 10*

SCORE	/40
mins	secs

TARGET(S)

How did you do this time ?

What can you do to improve ?

Take the challenge – You can do it !

$6\overline{)36}$	$7\overline{)21}$	$9\overline{)54}$	$4\overline{)16}$	$5\overline{)50}$
$5\overline{)35}$	$4\overline{)12}$	$8\overline{)72}$	$10\overline{)30}$	$8\overline{)24}$
$3\overline{)18}$	$9\overline{)81}$	$4\overline{)28}$	$5\overline{)10}$	$8\overline{)32}$
$5\overline{)25}$	$4\overline{)40}$	$2\overline{)16}$	$7\overline{)49}$	$9\overline{)90}$
$8\overline{)56}$	$3\overline{)9}$	$10\overline{)60}$	$5\overline{)40}$	$9\overline{)18}$
$3\overline{)15}$	$8\overline{)64}$	$6\overline{)24}$	$4\overline{)36}$	$5\overline{)30}$
$9\overline{)45}$	$4\overline{)20}$	$8\overline{)80}$	$2\overline{)12}$	$6\overline{)48}$
$10\overline{)70}$	$7\overline{)14}$	$9\overline{)63}$	$3\overline{)27}$	$7\overline{)42}$

NAME: _____   DATE: _____

THE DIVIDING WALL

KEY FOCUS: *Mentally dividing by 2 to 10*

Extension 1c

SCORE	/40
mins	secs

TARGET(S)

How did you do this time ?

What can you do to improve ?

Take the challenge – You can do it !

3 ⟌ 2 4	6 ⟌ 4 2	5 ⟌ 2 5	2 ⟌ 1 4	7 ⟌ 7 0
6 ⟌ 3 0	3 ⟌ 1 2	10 ⟌ 5 0	9 ⟌ 2 7	4 ⟌ 3 2
8 ⟌ 4 8	5 ⟌ 2 0	7 ⟌ 6 3	6 ⟌ 1 2	10 ⟌ 9 0
2 ⟌ 1 0	7 ⟌ 3 5	8 ⟌ 1 6	5 ⟌ 4 5	9 ⟌ 3 6
8 ⟌ 4 0	2 ⟌ 1 8	3 ⟌ 3 0	6 ⟌ 5 4	5 ⟌ 1 5
7 ⟌ 2 8	6 ⟌ 3 6	9 ⟌ 8 1	3 ⟌ 9	7 ⟌ 5 6
4 ⟌ 1 6	9 ⟌ 7 2	4 ⟌ 2 4	7 ⟌ 4 9	10 ⟌ 8 0
8 ⟌ 6 4	3 ⟌ 2 1	10 ⟌ 4 0	6 ⟌ 1 8	6 ⟌ 6 0

NAME: _____   **DATE:** _____

THE DIVIDING WALL

KEY FOCUS: *Mentally dividing by 2 to 10, with remainders*

Extension **2a**

SCORE	/40
mins	secs

| TARGET(S) | |

How did you do this time ?

What can you do to improve ?

Take the challenge – You can do it !

3 ⟌ 1 0	5 ⟌ 1 2	2 ⟌ 1 3	4 ⟌ 1 1	3 ⟌ 1 4
7 ⟌ 1 5	3 ⟌ 1 7	6 ⟌ 1 6	8 ⟌ 1 8	5 ⟌ 2 1
2 ⟌ 1 9	9 ⟌ 2 2	3 ⟌ 2 0	6 ⟌ 2 5	4 ⟌ 2 3
8 ⟌ 2 8	3 ⟌ 2 6	7 ⟌ 2 7	9 ⟌ 3 0	6 ⟌ 3 2
5 ⟌ 3 1	4 ⟌ 2 9	10 ⟌ 3 6	7 ⟌ 3 3	8 ⟌ 3 8
4 ⟌ 3 7	7 ⟌ 4 0	5 ⟌ 3 9	4 ⟌ 3 4	6 ⟌ 4 1
8 ⟌ 4 6	9 ⟌ 4 2	7 ⟌ 4 7	10 ⟌ 4 3	8 ⟌ 5 3
6 ⟌ 5 0	10 ⟌ 5 7	9 ⟌ 4 9	5 ⟌ 4 4	9 ⟌ 5 5

NAME: _____   DATE: _____

THE DIVIDING WALL

KEY FOCUS: *Mentally dividing by 2 to 10, with remainders*

Extension 2b

SCORE	/ 40
	mins secs

TARGET(S)

How did you do this time ?

What can you do to improve ?

Take the challenge – You can do it !

7 ⟌ 2 4   3 ⟌ 1 9   5 ⟌ 3 7   8 ⟌ 2 1   6 ⟌ 3 1

5 ⟌ 4 1   4 ⟌ 2 2   7 ⟌ 3 0   3 ⟌ 1 3   9 ⟌ 3 9

2 ⟌ 1 1   8 ⟌ 3 5   6 ⟌ 2 0   10 ⟌ 5 2   5 ⟌ 2 3

5 ⟌ 3 2   9 ⟌ 6 5   4 ⟌ 1 5   7 ⟌ 3 6   8 ⟌ 4 9

4 ⟌ 3 8   3 ⟌ 2 5   5 ⟌ 4 8   6 ⟌ 4 0   2 ⟌ 1 7

10 ⟌ 5 9   8 ⟌ 4 2   4 ⟌ 2 7   9 ⟌ 5 7   6 ⟌ 5 3

9 ⟌ 2 6   10 ⟌ 6 8   6 ⟌ 4 5   7 ⟌ 5 4   4 ⟌ 3 3

7 ⟌ 4 4   8 ⟌ 5 8   7 ⟌ 6 1   3 ⟌ 2 9   9 ⟌ 4 7

NAME: _____   DATE: _____

THE DIVIDING WALL

Extension 2c

KEY FOCUS: *Mentally dividing by 2 to 10, with remainders*

SCORE	/ 40
mins	secs
TARGET(S)	

How did you do this time ?

What can you do to improve ?

Take the challenge - You can do it !

4 ⟌ 3 5	3 ⟌ 2 3	7 ⟌ 2 5	5 ⟌ 3 8	9 ⟌ 3 4
6 ⟌ 2 9	8 ⟌ 5 2	5 ⟌ 2 4	7 ⟌ 6 2	2 ⟌ 1 5
4 ⟌ 1 7	9 ⟌ 4 0	5 ⟌ 3 3	9 ⟌ 7 5	8 ⟌ 3 6
5 ⟌ 4 3	6 ⟌ 3 7	8 ⟌ 7 3	4 ⟌ 2 6	9 ⟌ 5 3
8 ⟌ 5 9	3 ⟌ 1 6	5 ⟌ 4 6	7 ⟌ 4 1	10 ⟌ 7 7
6 ⟌ 5 1	4 ⟌ 3 1	9 ⟌ 4 8	8 ⟌ 6 7	3 ⟌ 2 8
7 ⟌ 4 5	6 ⟌ 5 6	7 ⟌ 3 2	4 ⟌ 2 1	10 ⟌ 6 3
8 ⟌ 2 7	7 ⟌ 5 5	4 ⟌ 3 9	9 ⟌ 6 6	6 ⟌ 4 7

NAME: _____   DATE: _____

THE DIVIDING WALL

Extension 2d

KEY FOCUS: *Mentally dividing by 2 to 10, with remainders*

How did you do this time ?

SCORE	/ 40
	mins secs
TARGET(S)	

What can you do to improve ?

Take the challenge – You can do it !

5 ⟌ 3 7	3 ⟌ 2 5	4 ⟌ 3 8	6 ⟌ 2 1	7 ⟌ 4 0
4 ⟌ 1 9	9 ⟌ 6 0	5 ⟌ 4 4	10 ⟌ 7 3	6 ⟌ 2 8
8 ⟌ 4 1	5 ⟌ 2 4	6 ⟌ 3 4	7 ⟌ 4 3	8 ⟌ 6 1
7 ⟌ 2 7	6 ⟌ 5 2	8 ⟌ 4 5	4 ⟌ 2 3	9 ⟌ 8 3
9 ⟌ 7 1	4 ⟌ 3 3	2 ⟌ 1 7	6 ⟌ 3 9	9 ⟌ 4 2
8 ⟌ 2 6	7 ⟌ 5 0	9 ⟌ 7 6	5 ⟌ 3 1	10 ⟌ 8 8
5 ⟌ 4 7	8 ⟌ 6 9	6 ⟌ 4 6	7 ⟌ 5 7	3 ⟌ 2 2
10 ⟌ 9 4	4 ⟌ 2 9	7 ⟌ 6 6	9 ⟌ 4 9	8 ⟌ 5 1

NAME: _____   DATE: _____

THE DIVIDING WALL

KEY FOCUS: *Mentally dividing by 2 to 10, with remainders*

Extension 2e

SCORE		/40
	mins	secs
TARGET(S)		

How did you do this time ?

What can you do to improve ?

Take the challenge – You can do it !

6 ⟌ 3 8	8 ⟌ 3 0	5 ⟌ 4 8	7 ⟌ 2 3	4 ⟌ 3 5
5 ⟌ 2 2	7 ⟌ 3 7	8 ⟌ 7 7	6 ⟌ 4 7	10 ⟌ 5 4
6 ⟌ 5 7	4 ⟌ 4 1	4 ⟌ 2 5	9 ⟌ 8 0	8 ⟌ 3 9
7 ⟌ 6 8	6 ⟌ 3 3	8 ⟌ 6 3	7 ⟌ 5 2	9 ⟌ 9 2
8 ⟌ 5 0	9 ⟌ 7 4	7 ⟌ 3 4	10 ⟌ 7 9	5 ⟌ 2 9
4 ⟌ 3 1	7 ⟌ 5 9	8 ⟌ 8 3	6 ⟌ 4 3	8 ⟌ 7 1
9 ⟌ 8 6	5 ⟌ 3 6	10 ⟌ 9 9	7 ⟌ 7 3	9 ⟌ 6 1
7 ⟌ 4 6	10 ⟌ 8 7	8 ⟌ 6 0	9 ⟌ 6 7	6 ⟌ 5 1

NAME: _____   DATE: _____

THE DIVIDING WALL

KEY FOCUS: *Dividing numbers up to 999 by 2 to 9*

Extension 3a

SCORE	/32
mins	secs

TARGET(S)

How did you do this time ?

What can you do to improve ?

Take the challenge – You can do it !

4) 2 9 2	3) 2 5 2	6) 2 1 6	5) 2 3 5
7) 4 3 4	2) 1 8 8	8) 3 0 4	4) 3 8 4
9) 5 3 1	3) 2 2 5	7) 3 7 1	2) 1 7 4
5) 3 3 5	6) 4 4 4	4) 3 3 2	9) 5 8 5
8) 9 7 6	7) 9 3 8	3) 8 2 8	5) 8 1 5
4) 9 1 6	6) 8 8 2	9) 9 2 7	7) 8 9 6
5) 9 3 0	2) 7 8 8	6) 8 1 6	4) 8 2 8
8) 8 6 4	3) 9 7 8	5) 7 4 5	6) 9 1 8

THE DIVIDING WALL

© 2015 Tony Colledge

KEY FOCUS: *Dividing numbers up to 999 by 2 to 9*

Extension **3b**

SCORE	/ 32
mins	secs

TARGET(S)

How did you do this time ?

What can you do to improve ?

Take the challenge – You can do it !

5) 2 9 5	4) 3 0 8	3) 2 4 9	6) 4 4 4
2) 2 7 2	7) 5 9 5	9) 5 7 6	3) 5 8 8
8) 5 3 6	4) 3 7 2	5) 4 3 5	6) 5 9 4
2) 5 3 8	7) 7 2 1	8) 5 9 2	4) 5 4 4
9) 7 3 8	3) 8 6 1	5) 6 3 0	6) 7 0 8
8) 7 0 4	2) 7 7 4	7) 8 8 2	4) 8 3 6
9) 8 7 3	3) 9 7 2	6) 7 3 8	5) 9 2 5
2) 8 7 6	8) 9 7 6	4) 9 3 6	7) 9 7 3

NAME: _____   DATE: _____

THE DIVIDING WALL

KEY FOCUS: *Dividing numbers up to 999 by 2 to 9*

Extension 3C

SCORE		/32
	mins	secs
TARGET(S)		

How did you do this time ?

What can you do to improve ?

Take the challenge – You can do it !

6) 3 8 4	3) 2 9 4	8) 5 7 6	7) 3 7 1
4) 3 4 4	9) 6 7 5	5) 3 3 5	6) 4 9 8
2) 4 5 4	7) 6 0 2	4) 5 1 2	8) 7 5 2
9) 7 4 7	5) 6 3 5	3) 7 9 5	6) 6 4 2
2) 7 9 6	4) 6 3 6	8) 8 4 0	7) 8 8 9
3) 9 5 7	9) 8 3 7	5) 8 3 0	6) 8 0 4
4) 8 2 4	8) 9 5 2	2) 9 2 6	7) 9 3 8
5) 9 6 5	4) 9 1 6	6) 9 4 2	9) 9 7 2

THE DIVIDING WALL

Extension 4a

KEY FOCUS: *Dividing numbers up to 9999 by 2 to 9*

NAME: _____   DATE: _____

SCORE /32

mins   secs

TARGET(S)

How did you do this time ?

What can you do to improve ?

Take the challenge – You can do it !

3) 1 8 8 4	5) 3 7 4 5	2) 6 8 1 4	6) 2 9 0 4
4) 2 6 6 8	7) 2 7 7 2	9) 3 1 8 6	8) 2 9 5 2
3) 2 5 1 1	6) 3 8 9 4	5) 6 1 8 0	4) 6 1 1 2
2) 4 1 4 6	8) 4 6 2 4	7) 5 0 7 5	9) 5 5 7 1
3) 5 0 7 9	5) 4 7 3 5	3) 7 5 2 4	6) 7 2 5 4
4) 5 4 7 6	9) 7 6 8 6	8) 7 8 6 4	2) 7 5 0 8
7) 6 6 3 6	5) 9 3 8 0	8) 9 6 5 6	3) 9 5 1 6
6) 9 4 9 2	4) 7 7 8 4	9) 8 7 5 7	7) 9 5 4 8

NAME: _____   **DATE:** _____

THE DIVIDING WALL

Extension 4b

KEY FOCUS: *Dividing numbers up to 9999 by 2 to 9*

How did you do this time ?

SCORE		/32
	mins	secs

What can you do to improve ?

TARGET(S)

Take the challenge - You can do it !

4) 2 9 1 2	6) 3 3 7 8	7) 3 4 0 9	5) 1 9 1 5
8) 2 3 8 4	3) 2 1 2 1	9) 2 1 5 1	2) 7 8 9 2
4) 2 4 3 2	3) 2 8 9 2	5) 3 9 2 5	6) 4 9 9 2
8) 5 4 7 2	7) 6 1 3 2	3) 4 1 5 8	9) 5 1 0 3
4) 3 7 7 2	8) 4 0 7 2	6) 8 2 6 8	7) 6 9 7 2
5) 5 3 7 5	2) 9 8 7 2	8) 6 6 1 6	4) 4 3 0 0
3) 8 6 8 2	7) 8 8 7 6	9) 8 6 8 5	5) 9 6 3 5
8) 8 7 6 8	6) 9 8 5 8	9) 9 9 7 2	4) 9 2 3 6

THE DIVIDING WALL

Extension 4C

KEY FOCUS: *Dividing numbers up to 9999 by 2 to 9*

How did you do this time ?

What can you do to improve ?

SCORE	/32
mins	secs
TARGET(S)	

Take the challenge – You can do it !

6) 2 3 8 8	5) 2 1 8 5	3) 2 5 7 7	8) 5 4 0 0
2) 3 3 5 6	7) 2 3 0 3	4) 2 7 3 2	9) 3 3 6 6
8) 6 4 4 8	6) 3 9 9 0	7) 3 8 2 9	5) 4 1 3 0
3) 4 3 9 5	4) 3 2 2 8	9) 5 6 3 4	7) 6 5 6 6
6) 7 7 1 6	9) 7 6 0 5	2) 7 0 1 2	5) 4 8 9 0
4) 7 4 1 6	8) 7 7 1 2	3) 9 8 0 1	7) 9 6 3 9
5) 8 8 2 0	4) 9 5 5 6	6) 9 4 4 4	9) 8 8 8 3
8) 9 9 0 4	2) 9 4 7 8	7) 7 6 5 1	5) 9 6 7 0

NAME: _____   DATE: _____

THE
DIVIDING WALL

Extension 5a

© 2015 Tony Colledge   **KEY FOCUS**: *Dividing large numbers by 2 to 9*

How did you do this time ?

SCORE		/20
	mins	secs
TARGET(S)		

What can you do to improve ?

Can you complete this harder challenge ?

3 | 7 9 4 2 7 4   5 | 8 6 9 4 7 0   4 | 7 7 1 0 5 2

6 | 5 7 2 3 1 6   8 | 9 9 0 1 7 6   7 | 9 7 0 5 2 9

9 | 5 4 8 4 2 4   2 | 7 5 1 8 9 6   6 | 4 9 6 8 1 8

4 | 6 8 2 6 3 6   7 | 6 4 8 0 1 8   8 | 7 0 1 0 9 6

5 | 2 3 1 2 9 9 0   3 | 8 8 4 2 5 4 9 2

9 | 6 5 7 5 9 3 0 9 1   4 | 6 6 5 2 8 3 0 3 2

7 | 4 3 8 0 6 6 5 4 2 5 5   6 | 5 2 9 4 2 5 0 8 6 7 6

5 | 3 7 3 5 1 4 3 1 9 6 9 0   8 | 9 8 9 0 1 6 4 7 7 5 3 6

THE DIVIDING WALL

Extension 5b

KEY FOCUS: *Dividing large numbers by 2 to 9*

How did you do this time ?

What can you do to improve ?

SCORE	/20
mins	secs
TARGET(S)	

Can you complete this harder challenge ?

$$4 \overline{)335704} \qquad 5 \overline{)337745} \qquad 7 \overline{)340879}$$

$$6 \overline{)539016} \qquad 8 \overline{)582864} \qquad 3 \overline{)811941}$$

$$9 \overline{)752166} \qquad 2 \overline{)892778} \qquad 4 \overline{)669624}$$

$$7 \overline{)424781} \qquad 5 \overline{)479125} \qquad 3 \overline{)882114}$$

$$6 \overline{)4501704} \qquad 9 \overline{)5253354}$$

$$8 \overline{)287217416} \qquad 5 \overline{)281910985}$$

$$3 \overline{)12781487052} \qquad 4 \overline{)37675101532}$$

$$7 \overline{)574535453653} \qquad 6 \overline{)949966049796}$$

The Dividing Wall Extension

ANSWERS

ANSWERS

THE DIVIDING WALL

Extension 1b

Total 40

© 2015 Tony Colledge

What was the final score ?

Were the previous targets beaten ?

What are the targets for next time ?

Keep trying hard – You can do it !

Divisor) Dividend	=	Divisor) Dividend	=	Divisor) Dividend	=	Divisor) Dividend	=	Divisor) Dividend	=
6) 36	6	7) 21	3	9) 54	6	4) 16	4	5) 50	10
5) 35	7	4) 12	3	8) 72	9	10) 30	3	8) 24	3
3) 18	6	9) 81	9	4) 28	7	5) 10	2	8) 32	4
5) 25	5	4) 40	10	2) 16	8	7) 49	7	9) 90	10
8) 56	7	3) 9	3	10) 60	6	5) 40	8	9) 18	2
3) 15	5	8) 64	8	6) 24	4	4) 36	9	5) 30	6
9) 45	5	4) 20	5	8) 80	10	2) 12	6	6) 48	8
10) 70	7	7) 14	2	9) 63	7	3) 27	9	7) 42	6

ANSWERS

THE DIVIDING WALL

Extension 1a

Total 40

© 2015 Tony Colledge

What was the final score ?

Were the previous targets beaten ?

What are the targets for next time ?

Keep trying hard – You can do it !

Divisor) Dividend	=	Divisor) Dividend	=	Divisor) Dividend	=	Divisor) Dividend	=	Divisor) Dividend	=
2) 12	6	5) 10	2	4) 16	4	3) 12	4	5) 15	3
7) 14	2	2) 16	8	5) 25	5	9) 18	2	4) 24	6
3) 9	3	6) 18	3	8) 32	4	4) 20	5	3) 27	9
3) 21	7	8) 24	3	5) 30	6	4) 36	9	5) 35	7
7) 28	4	3) 30	10	9) 54	6	8) 40	5	6) 42	7
6) 36	6	9) 45	5	10) 50	5	7) 49	7	9) 72	8
4) 40	10	7) 70	10	8) 48	6	10) 80	8	7) 63	9
9) 81	9	7) 56	8	6) 60	10	8) 64	8	10) 90	9

THE DIVIDING WALL — ANSWERS

Extension 2a

Total 40

What was the final score ?

Were the previous targets beaten ?

What are the targets for next time ?

Keep trying hard – You can do it !

10 ÷ 3 = 3 r1	12 ÷ 5 = 2 r2	13 ÷ 2 = 6 r1	11 ÷ 4 = 2 r3	14 ÷ 3 = 4 r2
15 ÷ 7 = 2 r1	17 ÷ 3 = 5 r2	16 ÷ 7 = 2 r2	18 ÷ 8 = 2 r2	21 ÷ 5 = 4 r1
19 ÷ 2 = 9 r1	22 ÷ 9 = 2 r4	20 ÷ 3 = 6 r2	25 ÷ 6 = 4 r1	23 ÷ 4 = 5 r3
28 ÷ 8 = 3 r4	26 ÷ 3 = 8 r2	27 ÷ 7 = 3 r6	30 ÷ 9 = 3 r3	32 ÷ 6 = 5 r2
31 ÷ 5 = 6 r1	29 ÷ 4 = 7 r1	36 ÷ 10 = 3 r6	33 ÷ 7 = 4 r5	38 ÷ 8 = 4 r6
37 ÷ 4 = 9 r1	40 ÷ 7 = 5 r5	39 ÷ 5 = 7 r4	34 ÷ 4 = 8 r2	41 ÷ 6 = 6 r5
46 ÷ 8 = 5 r6	42 ÷ 9 = 4 r6	47 ÷ 7 = 6 r5	43 ÷ 10 = 4 r3	53 ÷ 8 = 6 r5
50 ÷ 6 = 8 r2	57 ÷ 10 = 5 r7	49 ÷ 9 = 5 r4	44 ÷ 5 = 8 r4	55 ÷ 9 = 6 r1

THE DIVIDING WALL — ANSWERS

Extension 1c

Total 40

What was the final score ?

Were the previous targets beaten ?

What are the targets for next time ?

Keep trying hard – You can do it !

24 ÷ 3 = 8	42 ÷ 6 = 7	25 ÷ 5 = 5	14 ÷ 7 = 2	70 ÷ 7 = 10
30 ÷ 6 = 5	12 ÷ 3 = 4	50 ÷ 10 = 5	27 ÷ 9 = 3	32 ÷ 4 = 8
48 ÷ 8 = 6	20 ÷ 5 = 4	63 ÷ 7 = 9	12 ÷ 6 = 2	90 ÷ 10 = 9
10 ÷ 2 = 5	35 ÷ 7 = 5	16 ÷ 8 = 2	45 ÷ 5 = 9	36 ÷ 9 = 4
40 ÷ 8 = 5	18 ÷ 2 = 9	30 ÷ 3 = 10	54 ÷ 6 = 9	15 ÷ 5 = 3
28 ÷ 7 = 4	36 ÷ 6 = 6	81 ÷ 9 = 9	9 ÷ 3 = 3	56 ÷ 7 = 8
16 ÷ 4 = 4	72 ÷ 9 = 8	24 ÷ 4 = 6	49 ÷ 7 = 7	80 ÷ 10 = 8
64 ÷ 8 = 8	21 ÷ 3 = 7	40 ÷ 10 = 4	18 ÷ 6 = 3	60 ÷ 6 = 10

ANSWERS
THE DIVIDING WALL

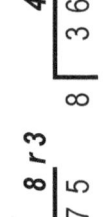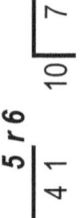

Extension 2c — Total 40

What was the final score ?

Were the previous targets beaten ?

What are the targets for next time ?

Keep trying hard – You can do it !

4) 35	8 r 3
3) 23	7 r 2
7) 25	3 r 4
5) 38	7 r 3
9) 34	3 r 7
6) 29	4 r 5
8) 52	6 r 4
5) 24	4 r 4
7) 62	8 r 6
2) 15	7 r 1
4) 17	4 r 1
9) 40	4 r 4
5) 33	6 r 3
9) 75	8 r 3
8) 36	4 r 4
5) 43	8 r 3
6) 37	6 r 1
8) 73	9 r 1
4) 26	6 r 2
9) 53	5 r 8
8) 59	7 r 3
3) 16	5 r 1
5) 46	9 r 1
7) 41	5 r 6
10) 77	7 r 7
6) 51	8 r 3
4) 31	7 r 3
9) 48	5 r 3
8) 67	8 r 3
3) 28	9 r 1
7) 45	6 r 3
7) 56	9 r 2
7) 32	4 r 4
4) 21	5 r 1
10) 63	6 r 3
8) 27	3 r 3
7) 55	7 r 6
4) 39	9 r 3
9) 66	7 r 3
6) 47	7 r 5

ANSWERS
THE DIVIDING WALL

Extension 2b — Total 40

What was the final score ?

Were the previous targets beaten ?

What are the targets for next time ?

Keep trying hard – You can do it !

3) 19	6 r 1
5) 37	7 r 2
6) 31	5 r 1
4) 22	5 r 2
7) 30	4 r 2
9) 39	4 r 3
8) 35	4 r 3
6) 20	3 r 2
5) 23	4 r 3
9) 65	7 r 2
4) 15	3 r 3
8) 49	6 r 1
8) 21	2 r 5
3) 25	8 r 1
5) 48	9 r 3
2) 17	8 r 1
3) 13	4 r 1
8) 42	5 r 2
4) 27	6 r 3
6) 53	8 r 5
10) 52	5 r 2
10) 68	6 r 8
6) 45	7 r 3
4) 33	8 r 1
7) 36	5 r 1
8) 58	7 r 2
7) 61	8 r 5
9) 47	5 r 2
7) 24	3 r 3
3) 19	6 r 1
6) 40	6 r 4
9) 57	6 r 3
5) 41	8 r 1
2) 11	5 r 1
5) 32	6 r 2
4) 38	9 r 2
10) 59	5 r 9
9) 26	2 r 8
7) 44	6 r 2
7) 54	7 r 5
3) 29	9 r 2

ANSWERS

THE DIVIDING WALL

Extension 2e

Total 40

What was the final score ?

Were the previous targets beaten ?

What are the targets for next time ?

Keep trying hard – You can do it !

8 r 3 4) 35	3 r 2 7) 23	9 r 3 5) 48	6 r 2 6) 38
5 r 4 10) 54	7 r 5 6) 47	9 r 5 8) 77	4 r 2 5) 22
4 r 7 8) 39	8 r 8 9) 80	6 r 1 4) 25	9 r 3 6) 57
10 r 2 9) 92	7 r 3 7) 52	7 r 7 8) 63	9 r 5 7) 68
5 r 4 5) 29	7 r 9 10) 79	4 r 6 7) 34	6 r 2 8) 50
8 r 7 8) 71	7 r 1 6) 43	10 r 3 8) 83	7 r 3 4) 31
6 r 7 9) 61	10 r 3 9) 73	9 r 9 10) 99	9 r 5 5) 86
8 r 3 6) 51	7 r 4 9) 67	7 r 4 8) 60	6 r 4 7) 46

ANSWERS

THE DIVIDING WALL

Extension 2d

Total 40

What was the final score ?

Were the previous targets beaten ?

What are the targets for next time ?

Keep trying hard – You can do it !

5 r 5 7) 40	3 r 3 6) 21	9 r 2 4) 38	8 r 1 3) 25
4 r 4 6) 28	7 r 3 10) 73	8 r 4 5) 44	6 r 6 9) 60
7 r 5 8) 61	6 r 1 7) 43	5 r 4 6) 34	4 r 4 5) 24
9 r 2 9) 83	5 r 3 4) 23	5 r 5 8) 45	8 r 4 6) 52
4 r 6 9) 42	6 r 3 6) 39	8 r 1 2) 17	8 r 1 4) 33
8 r 8 10) 88	6 r 1 5) 31	8 r 4 9) 76	7 r 1 7) 50
7 r 1 3) 22	8 r 1 7) 57	7 r 4 6) 46	8 r 5 8) 69
6 r 3 8) 51	5 r 4 9) 49	9 r 3 7) 66	7 r 1 4) 29
7 r 2 5) 37	4 r 3 4) 19	5 r 1 8) 41	3 r 6 7) 27
7 r 8 9) 71	3 r 2 8) 26	9 r 2 5) 47	9 r 4 10) 94

ANSWERS
THE DIVIDING WALL

Extension 3b — Total 32

What was the final score ?

Were the previous targets beaten ?

What are the targets for next time ?

Keep trying hard – You can do it !

59 5)295	77 4)308	83 3)249	74 6)444
136 2)272	85 7)595	64 9)576	196 3)588
67 8)536	93 4)372	87 5)435	99 6)594
269 2)538	103 7)721	74 8)592	136 4)544
82 9)738	287 3)861	126 5)630	118 6)708
88 8)704	387 2)774	126 7)882	209 4)836
97 9)873	324 3)972	123 6)738	185 5)925
438 2)876	122 8)976	234 4)936	139 7)973

ANSWERS
THE DIVIDING WALL

Extension 3a — Total 32

What was the final score ?

Were the previous targets beaten ?

What are the targets for next time ?

Keep trying hard – You can do it !

73 4)292	84 3)252	36 6)216	47 5)235
62 7)434	94 2)188	38 8)304	96 4)384
59 9)531	75 3)225	53 7)371	87 2)174
67 5)335	74 6)444	83 4)332	65 9)585
122 8)976	134 7)938	276 3)828	163 5)815
229 4)916	147 6)882	103 9)927	128 7)896
186 5)930	394 2)788	136 6)816	207 4)828
108 8)864	326 3)978	149 5)745	153 6)918

ANSWERS

THE DIVIDING WALL

Extension 4a

Total 32

© 2015 Tony Colledge

What was the final score ?

Were the previous targets beaten ?

What are the targets for next time ?

Keep trying hard – You can do it !

| 628 | 749 | 3407 | 484 |
| 3)1884 | 5)3745 | 2)6814 | 6)2904 |

| 667 | 396 | 354 | 369 |
| 4)2668 | 7)2772 | 9)3186 | 8)2952 |

| 837 | 649 | 1236 | 1528 |
| 3)2511 | 6)3894 | 5)6180 | 4)6112 |

| 2073 | 578 | 725 | 619 |
| 2)4146 | 8)4624 | 7)5075 | 9)5571 |

| 1693 | 947 | 2508 | 1209 |
| 3)5079 | 5)4735 | 3)7524 | 6)7254 |

| 1369 | 854 | 983 | 3754 |
| 4)5476 | 9)7686 | 8)7864 | 2)7508 |

| 948 | 1876 | 1207 | 3172 |
| 7)6636 | 5)9380 | 8)9656 | 3)9516 |

| 1582 | 1946 | 973 | 1364 |
| 6)9492 | 4)7784 | 9)8757 | 7)9548 |

ANSWERS

THE DIVIDING WALL

Extension 3c

Total 32

© 2015 Tony Colledge

What was the final score ?

Were the previous targets beaten ?

What are the targets for next time ?

Keep trying hard – You can do it !

| 64 | 98 | 72 | 53 |
| 6)384 | 3)294 | 8)576 | 7)371 |

| 86 | 75 | 67 | 83 |
| 4)344 | 9)675 | 5)335 | 6)498 |

| 227 | 86 | 128 | 94 |
| 2)454 | 7)602 | 4)512 | 8)752 |

| 83 | 127 | 265 | 107 |
| 9)747 | 5)635 | 3)795 | 6)642 |

| 398 | 159 | 105 | 127 |
| 2)796 | 4)636 | 8)840 | 7)889 |

| 319 | 93 | 166 | 134 |
| 3)957 | 9)837 | 5)830 | 6)804 |

| 206 | 119 | 463 | 134 |
| 4)824 | 8)952 | 2)926 | 7)938 |

| 193 | 229 | 157 | 108 |
| 5)965 | 4)916 | 6)942 | 9)972 |

ANSWERS

THE DIVIDING WALL

Extension 4c

Total 32

What was the final score ?

Were the previous targets beaten ?

What are the targets for next time ?

Keep trying hard – You can do it !

398 — 6)2388	437 — 5)2185	859 — 3)2577	675 — 8)5400
1678 — 2)3356	329 — 7)2303	683 — 4)2732	374 — 9)3366
806 — 8)6448	665 — 6)3990	547 — 4)3829	826 — 9)4130
1465 — 3)4395	807 — 4)3228	626 — 9)5634	938 — 7)6566
1286 — 6)7716	845 — 9)7605	3506 — 2)7012	978 — 5)4890
1854 — 4)7416	964 — 8)7712	3267 — 3)9801	1377 — 7)9639
1764 — 5)8820	2389 — 4)9556	1574 — 6)9444	987 — 9)8883
1238 — 8)9904	4739 — 2)9478	1093 — 7)7651	1934 — 5)9670

ANSWERS

THE DIVIDING WALL

Extension 4b

Total 32

What was the final score ?

Were the previous targets beaten ?

What are the targets for next time ?

Keep trying hard – You can do it !

728 — 4)2912	563 — 6)3378	487 — 7)3409	383 — 5)1915
298 — 8)2384	707 — 3)2121	239 — 9)2151	3946 — 2)7892
608 — 4)2432	964 — 3)2892	785 — 5)3925	832 — 6)4992
684 — 8)5472	876 — 7)6132	1386 — 3)4158	567 — 9)5103
943 — 4)3772	509 — 8)4072	1378 — 6)8268	996 — 7)6972
1075 — 5)5375	4936 — 2)9872	827 — 8)6616	1075 — 4)4300
2894 — 3)8682	1268 — 7)8876	965 — 9)8685	1927 — 5)9635
1096 — 8)8768	1643 — 6)9858	1108 — 9)9972	2309 — 4)9236

ANSWERS
THE DIVIDING WALL

Extension 5b
Total 20

What was the final score ?

Were the previous targets beaten ?

What are the targets for next time ?

Keep trying hard – You can do it !

```
  83926        67549        48697
4)335704     5)337745     7)340879

  89836        72858       270647
6)539016     8)582864     3)811941

  83574       446389      167406
9)752166     2)892778     4)669624

  60683        95825       294038
7)424781     5)479125     3)882114

   750284        583706
6)4501704     9)5253354

   35902177        56382197
8)287217416     5)281910985

    4260495684          9418775383
3)12781487052      4)37675101532

    82076493379         158327674966
7)574535453653     6)949966049796
```

ANSWERS
THE DIVIDING WALL

Extension 5a
Total 20

What was the final score ?

Were the previous targets beaten ?

What are the targets for next time ?

Keep trying hard – You can do it !

```
  264758       173894      192763
3)794274     5)869470     4)771052

  95386        123772      138647
6)572316     8)990176     7)970529

  60936        375948       82803
9)548424     2)751896     6)496818

  170659        92574       87637
4)682636     7)648018     8)701096

   462598          29475164
5)2312990       3)88425492

   73065899          166320758
9)657593091       4)665283032

    6258093465         8823751446
7)43806654255     6)52942508676

    74702863938          123627059692
5)373514319690     8)988016477536
```